Chevy Speed Manual

by FRED W. "BILL" FISHER

Contents

Introduction 2	Flywheel Chopping 80
Chevy Manual 3	How Far Shall I Go? 81
Dynamometer Tests 5	How To Rework Your GMC Engine 83
How To Build Your Chevrolet Engine ... 19	Lower End — Cylinder Head — Pistons —
Lower End—Firing It Up—Planning	Pushrods and Valve Tappets — Timing
Equipping Your Chevy With A Racing Cam.. 25	Gears & Cam Timing — Water Pump and
How To Select A Cam 28	Crank Pulley — Starter and Generator —
Semi—¾ Grind—Full Race—Super—Track	Clutch — Installing GMC Engines In Chevrolets
Cam Functioning 31	GMC Set Ups 95
Hom To Install A Racing Cam	GMC Hot Road Job—All-Out Racing Engine
Valve Components 35	— Assembly Differences — Winfield GMC
Valves — Valve Springs — Spring Height —	A Hudson-Buick 99
Spring Retainers — Rocker Arms — Rocker	How To Rebuild Your Buick For Speed 101
Arm Stands — Pushrods — Tappets — Rocker	Lower End — Pistons — Camshaft — Carburetion — Ignition — Cylinder Head — Exhaust —
Arm Covers	Flywheel and Clutch — Rocker Arms and
Pistons & Piston Rings 42	Pushrods
Special Heads 47	Tuning Your Engine 107
How To Rework Your Chevy Head ... 49	Dual Manifold Installation — Improper
Milling—Porting—Powerglide and 1950-51	Functioning — Improper Idling — Carburetors — Road Tuning
Hi-Torque Heads—Finishing—Seat Details	Exhaust Systems 114
How To Compute Compression Ratios ... 55	Fuel 116
Ignition 59	Water Injection 119
Special Dual-Point Ignition	Driving Your Hopped-Up Car 121
Caution! — Blasting! Read With Care 65	Trouble 122
No Mikes? 68	Engine Trouble—Determining The Trouble
1950 105-H.P. Hi-Torques 69	GMC Horsepower Outputs 127
Powerglide Chevys 70	
Chevy Pressure Oiling 74	
Gear Ratio 76	

P.O. Box 91858
Tucson, AZ 85752-1858
(520) 547-2462

Printed in U.S.A.
10 9 8 7 6 5 4

Introduction

Once you've seen, driven, or ridden in a really good going Chevrolet that has been given the speed treatment, you'll never again doubt that a Chevy can be made to develop lots and lots of usable horsepower.

The author, himself, scoffed at the "Cast-Iron Six" several years ago, being prone to feel that the V-8's were the only "Go-Fast" engines. However, one ride in a hot Chevrolet coupe was the convincing argument that made him "see the light." Please don't take this in the wrong way! Anyone who thinks that one particular engine is the "only one" for hotrodding is sooner or later going to get "blown-off" by another make. That there are many more "going" V-8's than Chevrolets cannot be denied, nor would it be reasonable to assume that any other engine will soon hold as many track, lakes, marine and drag racing records. This is undoubtedly due to the fact that V-8's and other Ford engines have been for years by far the most popular make of engine for souping-up.

It would also be unwise to assume that the Chevrolet and G. M. C. engines are not going to become more and more popular for all kinds of racing, due to the fine horsepower output which is available from these engines when properly modified.

We feel that our past two Chevrolet Speed Manuals have done much to bring this about. Certainly the manufacturers of speed equipment have seen the needs of the Chevrolet speed enthusiasts. A few years ago when the writer began to hop-up Chevrolet engines, not very much information was available to the general public, and special parts were almost unobtainable. Now, as we enter 1951, parts are widely available and there are many makes from which to choose. Now, as in the V-8 field, it is necessary to choose your equipment wisely so as to be assured that you are going to get the best performance for your money.

As you read on, you will see more clearly what can be accomplished by careful choice and installation of speed equipment in your Chevrolet engine.

Above all, the reader is advised to be meticulously careful in installing or reworking any part. Each bit of information in this manual has been placed here for your guidance. Much of it has been learned through expensively obtained experience. To avoid costly blow-ups, stick close to this data and you shall be rewarded with a fast, faithful, and sweet running engine.

Chevy Manual

Here is another speed manual for Chevrolets! The first two served their purpose—to get the country to realize that the V-8 engine was not the only one when it came to putting out usable horsepower for racing purposes.

Since the publication of the first 54-page Chevrolet speed Manual with the green cover, we have been engaged in producing hotter and hotter Chevrolet Engines for road, marine, and all-out lakes or track use. With the development of new items for improving the horsepower output of the Chevrolet engine, we have tried to test these in actual competition so that we could give our readers the best and latest information. In some cases this has caused us blow-ups of expensive engines, but we feel that this too, has been worth while, in that it has saved our customers from making the same mistakes.

Some of the items mentioned and given a "boost" in our first manual and in the second one, have not proven to be the best. Of course, newer parts and know-how have gone into the making of what is available today, until it is now possible to buy Chevrolet speed parts of every type and description right off of the counter just like the Ford owners have been able to do for many years.

As you build your Chevrolet engine, be careful to follow our instructions carefully. They are the product of years of experience and have been used to build hundreds of successful and fast Chevrolets.

Chevrolet owners, dealers, and speed shop owners come to our shop from all over the country to see the engines and parts which we have on display. Many are not really convinced that a Chevrolet can be made to run very fast. We point out that John Hartman's 154 mph roadster held the record at the Dry Lakes long after it was retired to run solely on the track. Marvin Lee's City of Pasadena Streamliner with the Chevrolet engine was clocked at 162 mph. Our 1937 maroon full-fendered Chevrolet coupe has run as high as 127 mph at the Russetta Timing Association meets.

If you must have larger displacement than can be obtained with the aluminum replacement type pistons, special solid skirt racing pistons are available, or you can go to the limit of cubic inches by installing a GMC truck motor, the installation of which is covered later on in the manual.

1941 GMC equipped Chevrolet owned and modified by Fred W. Fisher turns 92 mph. at quarter-mile drags, 126 mph. at dry lakes. Produces 170 h.p. at rear wheels on gasoline, and over 225 h.p. at rear wheels on alcohol fuel.

Some will protest that many of these records were made with the fabulous WAYNE equipment which is somewhat more expensive. We grant that this is true! However, the horsepower which can be produced from the Chevrolet engine with a stock head is also terrific, and many feel that with proper tuning the Chevrolet engine with a modified stock head can be made to run quite close to the performance of the very special competition heads.

It is obvious that the Over Head Valve type engines are more efficient (when properly tuned) than are flat head engines of comparable displacement. Not only that, but we have proven to ourselves that you can build a hot Chevrolet engine for less money than your friend, the V-8 owner, will have to spend.

Pound for pound, the Chevrolet engine is one of the most efficient on the market today, and for producing flashing speed, power, and performance for small displacement, few flat head engines can compare when horsepower per cubic inch is considered.

Dynamometer Tests

Most readers will probably wonder why we decided to use this semi-stock equipped Chevrolet engine for the dynamometer tests of which you are about to read. For almost a decade, perhaps longer, the speed-loving public has been deluged from all sides with claims of horsepower gains from this item of equipment, or another item of equipment; claims advertising horsepower increases which seem almost fantastic. In many cases these claims would be laughed at if it were known how much additional equipment were required to obtain this fabulous horsepower output.

For that reason we decided that it might be wise to use an engine which used semi-stock equipment, due to the fact that the average man wants to know just how much good his $30.00 or $100.00, or whatever he happens to want to spend, is going to do for his particular engine.

Naturally, we could have picked a larger Hi-Torque Chevrolet (truck) engine for these tests, as we had several available at the time. Most Chevrolet owners have Standard Engines; and used Hi-Torque block and crank assemblies at reasonable prices are always hard to obtain. Although it is generally known that a larger engine will produce more horse-power, due to displacement alone, we picked a Standard Chevrolet engine, bored out to 3⅝" with stock stroke of 3¾". Since all Chevrolet blocks, 1941 or later, can be bored out to this size, this is an engine which the average Chevrolet owner can duplicate with a minimum of expense. When this engine was set up it was line-bored to give a clearance of .0025" for the main bearings. In order that standard bearings might be used, the crank-shaft was ground to a standard undersize, plus an additional .001" to obtain the necessary clearance needed for high speed, high power out-put, or a total of .0035" for the connecting rod bearings.

Aluminum timing gears were used on each of the cams which we tried during the experiments. Several different cylinder head set-ups were used, as you will see from the dynamometer test reports. The pistons were the standard split-skirt, T-slot type aluminum pistons which we sell for Chevrolets for fast road use. These were fitted with .004" clearance on the skirts.

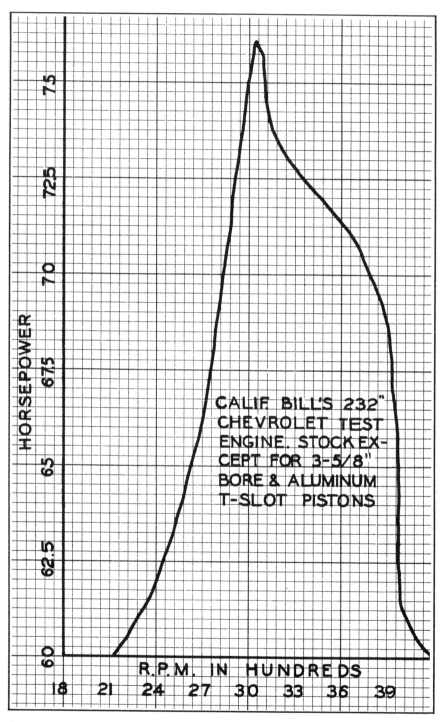

Special tubular push rods were used throughout the tests, along with Buick inner valve springs and Chevrolet outer valve springs. Insert rods were used with standard bearings, obtainable at any parts house. The stock Chevrolet oil pan was "targeted," in order to make sure that the oil spouts were directed toward the dippers on the bottom of the rods, as is standard Chevrolet practice. The height of the oil dippers was checked by means of special gauges which are available at your Chevrolet dealers. These two items, that is, the targeting of the pan and the oil dipper height, must be checked very closely, if continued successful high speed performance is wanted or desired from a Chevrolet engine with the stock oiling system.

When the engine was set up, new cam bearings were also installed, and each cam installed was carefully checked to make sure that it ran free in the cam bearings. Standard Chevrolet tappets, or cam followers were used in each test; and more information on valve followers or tappets is to be found under the section on valve components.

The author sees no reason why any reader, who carefully follows the instructions and directions contained in this Manual, should have any difficulty in building an engine comparable to that which we used for the dynamometer tests. We have used these engines on the street, in drag racing, and on the open road for competitive use for many thousands of miles, and they will stand up very well under continued abuse. As for lasting qualities, if one were to drive them just as a stock engine, the mileage obtained before overhaul would be equivalent to that which you could obtain from a stock Chevrolet factory motor.

It has been our intention to make this manual as complete in every way as is humanly possible. It has been checked and rechecked, in hope of finding any error or omission of any data which could be used profitably by the reader in building high-performance Chevrolet engines.

Here is the answer to speed tuning! Just get on an engine dyno with your outfit and find out what is slowing it down. We honestly had difficulty believing some of the things which we discovered on the dynamometer, and the time that we spent there was perhaps the most valuable that we have ever experienced.

To get the spark setting correct, the ignition can be changed with the engine running at full throttle to obtain the maximum output, cams can be changed in one hour flat with two fellows working on the engine. Mr. Nick Glaviano, of NICSON ENGINEERING, and the author quite easily changed from one cam to another in one hour. Manifold changes can be made in fifteen minutes or less, and perhaps the only change that we made that took a large amount of time was from the stock head to a high compression, ported head. That was largely due to the fact that we had to cut off the bottom of the stock side plate and slot the lower holes in this push rod cover panel.

Our objective in tuning up this stock engine was to find out just what each individual item of speed equipment would do for the Chevrolet engine, and what changes it would make in the points where peak horsepower was developed. Naturally we had already spent a lot of time tuning engines on chassis type dynamometers, but there it is impossible to make the changes which we were able to make on the engine dynamometer, where the engine is sitting right out in the open where you can easily get at any part of it.

Our first test was conducted with a Chevrolet engine, bored out to .060" over hi torque, and using insert rods. Clearances already have been described. A stock camshaft and stock head were used. Buick inner and outer valve springs with stock spring seats and keepers were used. One ROCHESTER carburetor used. Mobil Ethyl gasoline.

From this table it is easy to see that the peak horsepower with the stock Chevrolet engine is developed at 3300 RPM, and that the horsepower rapidly falls off above or below that figure. We were quite surprised that the engine did not put out the 90 horsepower as it is rated by the Chevrolet factory. It is not easy to account for a variance of 15 horsepower. We did all within our power to show more horsepower than this, but Mr. Johansen assured us that it is quite common for a stock engine to produce less than its rated horsepower. We felt better when he showed us results of his tests which showed that the 100 h. p. Ford V-8 produced less than the 90 h. p. Chevrolet.

After the stock test we installed a dual carburetor NICSON manifold with two ZENITH carburetors. Otherwise the set up was unchanged, except that we were able to advance the spark farther after the dual manifold was installed.

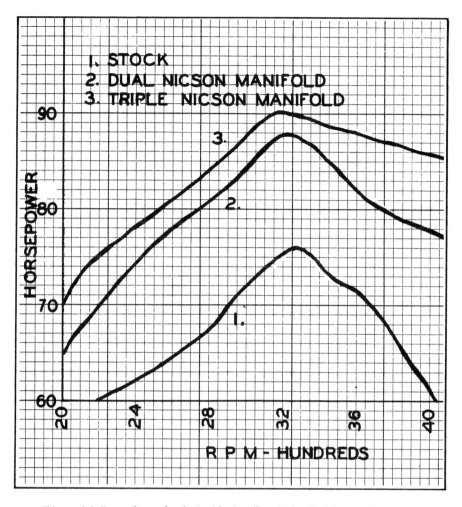

This graph indicates the results obtained by installing dual and triple manifolds on a stock Chevrolet engine. Although the difference between the two carb and three carb manifolds is only slight on a stock engine, the improvement provided by the three carb job is more apparent when other speed equipment is added to the engine.

WHAT A DIFFERENCE! We had a hard time believing it too, but there it is, just as the dynamometer told it to us! From 2000 RPM to 4000 RPM, an average gain of 12 horsepower! Look at that 4000 RPM figure, a full 16 horsepower greater than with the stock one carburetor manifold. Think of the improvement that would make for high speed cruising with effortless ease!

Next a three carburetor, down draft, NICSON triple manifold was installed. Rest of set up remained unchanged with exception of further advance on spark setting.

From these figures we can draw the following conclusions: (1) That the three carburetor manifold is definitely going to help the stock engine; (2) That the three carburetor manifold is superior to the dual; (3) That peak horsepower is still produced at 3200 RPM as with the dual. These three statements should be considered, as well as the fact that a careful study of the graph shows that the horsepower does not fall as rapidly at the upper RPM's with the triple as with the dual, thus making the horsepower curve flatten out.

Therefore, the three carburetor manifold is going to prove more advantageous for high speeds. Not only that, but the horsepower gains low down in the engine range will provide greater acceleration throughout the range of engine speeds.

Our next move was to remove the stock cam and install a ¾ RACE HOWARD camshaft. We were frankly quite surprised with the performance which we secured from this cam. Howard Johansen is constantly developing new and better cams and this one certainly shows a lot of time and thought, as well as careful experimentation in its development. The three carburetor manifold was removed and the stock manifold with a single ROCHESTER carburetor was installed. We had a bad time getting this combination to run, and it was obvious from the start that we were going to be undercarbureted with this cam and one carburetor. We made several attempts to get the mixture right with stock jets, with no success. We finally resorted to drilling out the main jet with a .0695" drill. This came pretty close to giving us the correct mixture, but we had no difficulty in discovering that the venturi of this carburetor was too small. It may be that the ROCHESTER, as supplied with the Power-Glide and 1950 105 h.p. hi-torques, will be large enough for use with this camshaft. Certainly a ZENITH with large venturi or a 1 - 7/32" STROMBERG BXOV-2 could be used to advantage.

AUTHOR'S NOTE: We have used the ¾ HOWARD cams on many cars with one carburetor and did not ever find the installation to be anything other than very helpful to the cars' performance. Apparently we were placing a much greater load on the engine with the dynamometer than is experienced in actual normal driving. However, we always notice a terrific improvement when dual carbs are installed where the engine has already been equipped with a cam.

At this point we decided that we were getting nowhere in a hurry, and therefore decided to see if the engine would run any better if we allowed it to rev up a bit more. We put the engine up to 4400 RPM under a full load and secured a good reading of 72.6 horsepower. This was the only reading that we secured with this combination that was not worse than we had with the stock set up. We were beginning to wonder if the cam were fouled up in some way, but when we put on the two carburetor manifold that engine purred like the cat in the creamery! Smooth idle, no tappet noise, and instant throttle response! From the minute we cracked the throttle on this set up it was obvious that we had a real combination! Above 3400 the exhaust note began to take on a ripping noise that was a delight to the ears!

The gains secured with this combination will satisfy a lot of motorists who are sick and tired of being left at 80 miles an hour by some larger, more expensive car. Our opinion would be that this set up on your standard Chevrolet engine, prepared as indicated in this book, will give you ten to fifteen miles per hour more top speed.

Peak horsepower is now obtained at 3600 RPM instead of 3300, and it is quite interesting to note that the low speed torque has not been at all impaired.

Off with the two jugger and on with a triple down draft. We were conducting this test just as we were getting ready to close down for the day. Due to our tight schedule, the head had to be off that evening so that we might examine the valves and the interior of the engine to make certain that the fuel mixtures were all right and that everything was in good condition. Not only that, but our further tests called for the installation of a different head the next day. Starting right in at 2200 RPM, we secured the following results.

RPM	HP	RPM	HP
2000	69.3	3200	89.6
2200	74.4	3400	88.4
2600	79.3	3600	88.2
2800	82.6	3800	87.4
3000	87.0	4000	80.0
		4200	75.6

GRAPH OF THESE RESULTS ON FOLLOWING PAGE

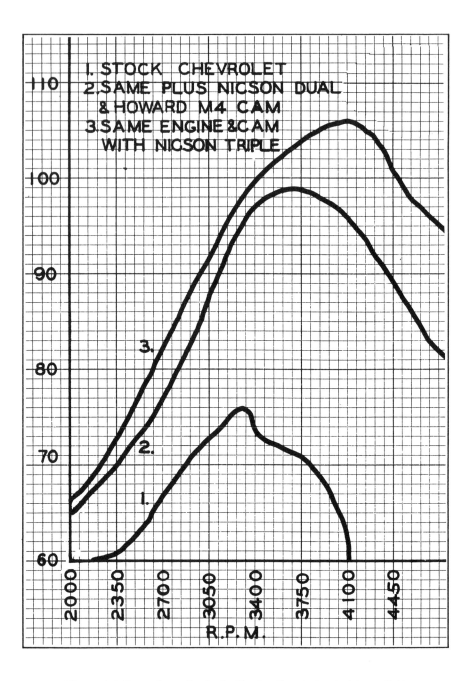

This graph indicates the results obtained by installing dual and triple manifolds on a stock Chevrolet engine. Although the difference between the two carb and three carb manifolds is only slight on a stock engine, the improvement provided by the three carb job is more apparent when other speed equipment is added to the engine.

At the conclusion of this test the red hot exhaust pipe and exhaust manifold showed beyond a doubt that the mixture was far too lean for the load under which we had been running the engine. However, no other jets were at hand, and it was obvious that we would have to enrich both the power and main jets for satisfactory operation. The J-3 Champion plugs which we had been using until that time testified to the fact that the last test had been run with too lean a mixture. We decided that the data which we had secured was conclusive enough and that to delay removing the head until the next day would not be advisable.

We promptly got busy and had the head off in fifteen minutes. A check of the exhaust valves showed that the chocolate brown or rust color that a proper mixture gives, was not apparent. Instead, they looked as if they had been sitting on the gates of Hell. Don't take this to mean that high speed and high power output is destructive to long valve life, as it is not so if the mixture is kept correct. It is always better to run just a bit rich than to have the engine constantly running under load with a lean fuel mixture!

As we planned to use some pretty radical cams for some of the future tests and we did not wish to remove the high compression head again to make valve spring changes, we installed BUICK outer with competition inner valve springs. This is not a combination we would recommend for the normal road job, due to the fact that the spring pressure tears up the valve tappets (cam followers) in a great hurry. Competition spring seats were used, with stock Chevrolet keepers and valves. The stock spring washers were used under the intake valve spring combinations, but these washers were omitted from the exhaust valve set ups.

On installing the head, all bolts were taken down to a 90 lb. reading with our SNAP-ON Torque-O-Meter wrench. Since this head was reworked from a 1937-40 casting, it had been milled 3/16" as well as having been filled. A check of the cranking compression with throttle held wide open and all spark plugs removed was 163 lbs./sq./in. Stock compression on the Chevrolet is supposed to be 110-112 lbs. at cranking speed, so this head should produce some real improvements in power as well as gasoline mileage. Let's take a look at what the dynamometer showed us.

We tried a muffler on this test at 44 and 46 hundred RPM's. Much to our disgust, the stock Chevrolet muffler was found to take 22 h.p.

What a horsepower robber! In our estimation, the stock Chevrolet muffler must be removed if performance is desired. Adequate exhaust systems are discussed in a separate chapter.

Our next move was to install an ISKENDERIAN "Bigelow" Track grind cam. This cam was named after Kenny Bigelow as he helped to pioneer its development. This cam is perhaps the noisiest we have ever heard when set with stock clearances, and as you will see by the following chart, it is least efficient when set with the recommended clearance of 20 and 22 thousandths. We found that setting the cam clearance with .006" and .012" produced far more horsepower. Mr. Bigelow happened to drop in at that time and gave us this clue, but cautioned users of this cam to run at least .015" exhaust clearance for normal road use, and about the same for the intake clearance. The closer tolerances can be used to advantage for all-out racing, due to the fact that they greatly increase the duration of this cam.

Valve bounce is a bad feature inherent in this particular grind. It definitely starts to fall off due to that fact above 4800 RPM. However, it does not lose horsepower as rapidly at high RPM's with the smaller clearances. We never recommend this cam for any Chevrolet without stressing the fact that WAYNE spring seats must be used on the intakes, and preferably on the exhausts. This camshaft is popular for stock racing where only one carburetor can be used according to the rules. Of course, a larger than stock carburetor must be used, preferably a Ford venturi.

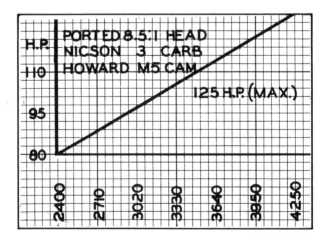

Three carburetor down draft manifold, ¾ Howard Camshaft, and 8.5:1 compression head on an otherwise stock Chevrolet engine. This is a terrific combination in any one's engine! This set up gives fifty more horsepower than stock, which is a real jump in developed horsepower.

Our next few days on the dynamometer were spent using special fuels. On some days we spent as much as $40.00 for fuel alone. The accompanying graph shows what we obtained using the engine with CLARK Headers, ISKENDERIAN Track Grind cam, BARKER Hi-Lift Rocker Arms, NICSON three carburetor manifold, and various fuels. The manifold was equipped with ZENITH Down Draft Carburetors.

For full throttle operation we found it necessary to go one step richer on the Power Jets, as you will notice in the data on installing dual and triple carburetors later in this book.

Testing this combination on straight gasoline (140 octane aviation gas in this instance) we found that the Power Jets were too small, as mentioned in the previous paragraph. About three horsepower was gained by improving the fuel mixture, and a further seven horses came out of the stable when 10% nitro methane with an equivalent amount of amyl acetate was added. The amyl acetate is added merely as a catalyst to make the nitro mix with the gasoline. Some aviation gasolines will mix with nitro without a mixing agent. In order to test whether they will mix, it is only necessary to pour a small quantity of the gasoline in a jar or graduate and add the amount of nitro (in percent) that you intend to use. If the two liquids are not miscible, the mixture will look queer, like oil on water, or water in gasoline.

A further test was run with 15 and 20% nitro which showed that this amount was too much, as the dyno readings fell way off due to severe detonation. Therefore, we recommend that no more than 10% nitro methane be added to gasoline fuel for use in a racing Chevrolet engine.

Now we switched over to the NICSON Three Carburetor Side Draft manifold, which is an old favorite of the author's and many others connected with Chevy racing installations. The carburetors on this manifold were modified for use with alcohol fuel, and the dyno readings promptly indicated 175 horsepower on straight methanol. Nitro methane was now added in varying percentages until we determined that 20% nitro methane with 10% water would give the best performance. More than this got us into detonation troubles again. Now our horsepower was getting close to 200, and we had reached the limit of the track cam, so we changed over to a HOWARD SU 9 A grind, which showed slightly more horsepower at the top end and seemed to have more torque all through the range. In fact, with this cam we pulled as much horsepower at 3000 RPM as a 3/8 x 3/8 Mercury full race motor.

For almost a whole day we experimented with adding hydrogen peroxide to the fuel. This did not add any horsepower at any time.

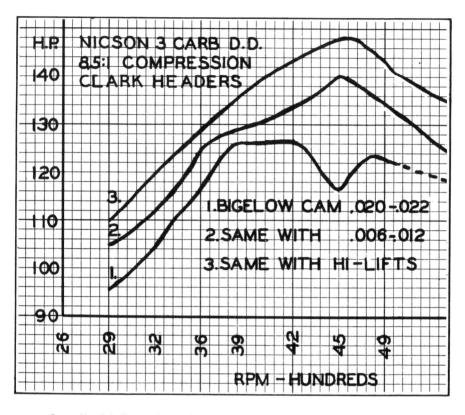

Curve No. 1 indicates the performance obtained from our test engine using an Iskenderian Bigelow Track Grind. This was installed using the recommended clearance, which was obviously too much. Since then, Iskenderian has changed the recommended clearances on this cam. We lessened the clearances to .006" and .012" to get the results shown by Curve No. 2. Curve No. 3 indicates the additional horsepower obtained when a set of Hi-Lift rocker arms was installed in conjunction with this Track Grind camshaft.

With the HOWARD SU 9A cam and NICSON Three Carb Side Draft manifold we got flash readings of well over 200 horses, and we feel it safe to assume that had the engine been a hi-torque with more inches and the longer stroke, that the engine would have produced 225 horsepower, using the same parts for the installation.

For the last day on the dynamometer we experimented with a special head (reworked stock Chevrolet). This head had intake valves from a Chevrolet Power Glide head, and stock Chevrolet intake valves for exhausts. The head did not provide any more horsepower at top speed, but did seem to have more torque through the engine range on the way up to top speeds. From this we can draw the conclusion that larger intake valves will aid the acceleration of most Chevrolets.

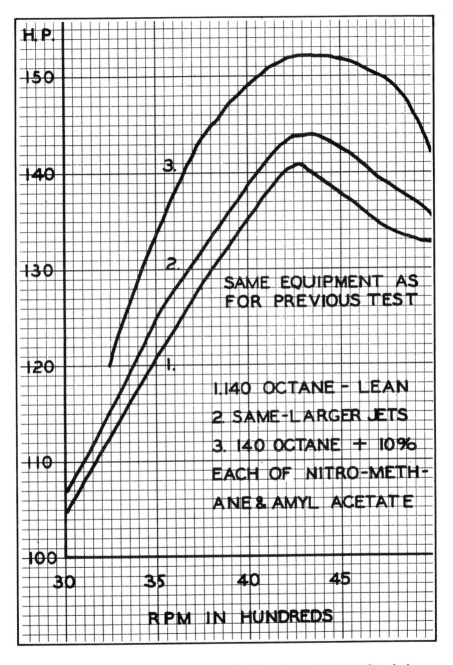

This graph indicates dynamometer test using 140 octane aviation gasoline, both straight and with nitro methane added. This fuel is more expensive than straight methanol, and will not provide the horsepower available from the alcohol type fuels. Alcohol fuel should be used whenever possible.

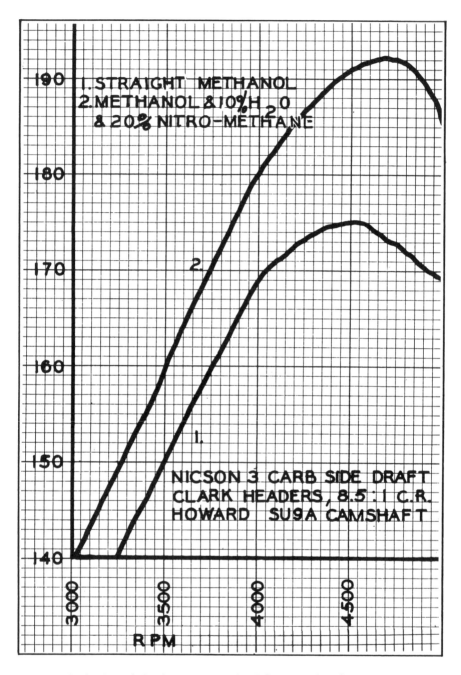

Graph showing relative horsepower produced by 232" Chevrolet engine using straight methanol fuel, with full race equipment, as compared with methanol with 20% nitro methane and 10% water. The water must be added to eliminate detonation.

How To Build Your Chevy Engine

Assuming that you have the block, crank, pistons, cam, timing gears, gaskets, etc.; we can begin to build your engine. Allow yourself plenty of time, and under no circumstances try to do the job over a week-end so that you can have the car running on Monday. It just isn't a good idea, because there will always be some little part, nut or bolt that you will need and be unable to obtain.

LOWER END

We always start to assemble our engines from the "bottom up" beginning by boiling out the block in a hot tank so that all of the water and oil passages will be thoroughly cleaned.

The block is then rebored to the size which we have previously determined. Front main cap timing cover bolt holes are now tapped with a ⅜-16 tap. After blowing out chips from the block, the main bearings are placed in position, with no shims, and the caps are tightened. The main bearings are then line-bored to give .0025" clearance. Naturally, the crank must be reground (if necessary) before the line boring operation is begun.

After the line-bore operation has been completed and all oil passages blown out thoroughly, we install new cam bearings. A new Welch plug is then installed at the rear cam bearing. New Welch plugs are also installed in each end of the main oil galley, unless the engine is to be equipped with a pressure bottom. In this case, pipe plugs are screwed into each end of the main oil galley as is explained in the data on converting your engine to pressure oiling.

Next, the rear main oil seals are carefully installed in the block and in the rear main cap. Install the front motor plate if it has been removed. Be sure to prick punch the fillet head screws so that there will be no chance of their vibrating loose while the engine is in service.

Oil the main bearings liberally and install the crankshaft in the main bearings. Use Aviation Form-A-Gasket PERMATEX type sealer where the rear and front main caps are joined to the block. This is to prevent oil leakage. Torque the main bearings screws down to 90 to 100 ft. lbs.

Insert the oil pump in its boss. Insert the pointed retainer screw in the boss and screw it in finger tight so that it seats in the corresponding indentation in the oil pump body. If this screw is tightened very much past finger tightness it will distort the body of the oil pump and cause the lower end of the distributor shaft to completely freeze up. Be sure to bring the jam nut down tightly with a wrench so that the oil pump retaining screw will not back out.

Install the oil line from the oil pump to the block. Make sure that there are no cracks in the line.

Pistons should now be assembled on the rods. If using Chevrolet rods, the wristpin should be held onto a drift or punch placed in a vise. This will place the tension of tightening the wrist pin clamp entirely upon the drift and wristpin instead of tending to cause the rod to twist out of alignment. This is important, so don't neglect it!

Tighten these clamping screws to 30 ft. lbs. torque. Move the rod back and forth in the piston to each edge to make sure that the edge of the wristpin does not protrude out of the piston.

Make sure that the rods and caps are properly numbered. If there are numbers on the pistons, these should coincide with those on the rods.

The split skirt of a T-Slot type piston should be placed away from the cam, the clamping bolt for the wristpin goes toward the cam, and the numbers on the rod go toward the cam.

Remove the nuts from the rods and the caps from the rods, being careful in removing the nuts so as not to twist the rods out of alignment.

Place the piston rings onto the pistons, checking first for ring groove depth as suggested in the data on PISTONS AND PISTON RINGS.

Stagger the gaps in the rings so that no two gaps coincide. No gaps should be placed so as to be in line with the wrist pin boss, as this may cause a compression leak.

Ring gap width should be carefully checked by placing the ring in the cylinder in which it is to be used and measuring the gap with a feeler gauge. The ring should be placed squarely in the bore by pushing it down carefully a slight distance with the head of the piston.

Oil the rings and grooves liberally and install the pistons in their proper cylinders, using a ring compressor and being careful not to break the rings. The head of the piston can be pushed in or tapped lightly with the handle of a hammer or a plastic mallet.

Place the inserts in the rod and rod caps. Tighten rod nuts 45 to 50 ft. lbs. torque. Tighten PAL-NUTS finger tight plus an additional one-half turn.

Place the Woodruff key in the crankshaft and smear the inside of the steel timing gear with white lead. Drive the gear on to the crank until it seats solidly.

Install the brass cam retainer plate on the cam, place a Woodruff key in the cam keyway, and press on an aluminum car gear, using a suitable press.

Make sure that the #1 piston is on T.D.C. Install the cam in its bearings, making sure that the marks on the crank gear and cam gear are lined up properly. Insert the two screws which secure the cam retaining plate. Tighten them securely, being careful not to strip the threads. The crank must be in this #1 T.D.C. position so that the cam will clear the crank throws.

Install a new front crankshaft timing cover seal in the timing cover. The feathered edge of the seal should point inward toward crank. Drill two threaded holes to clear ⅜-16 screws. Bolt on the timing cover. Drive the crankshaft pulley onto the crank (harmonic balancer).

The oil pan can now be installed. Gaskets can be either greased or coated with the Aviation Form-A-Gasket compound already mentioned. The oil pan must be targeted before its installation. The heights of the dippers on the rods should also be checked, as well as the height of the spouts, and the height of the troughs in the pan.

Check everything over before bolting on the pan. Did you remember the dippers? Oil line to pump in place? Shroud and screen on pump? All O.K.? Then bolt on the pan, and right the engine so that it is resting on the pan.

Place the tappets in the block after oiling them. Install the rocker arm oil line through the block. This line goes through the water jacket to the oil distributor on the left side of the block. Be careful not to crack or break this line when installing it.

Place head gasket on block. Install assembled head on block and tighten head bolts to 90 ft. lbs. torque. If the head has been milled, then spacer washers equal to the amount of mill should be installed under each and every head bolt.

Bring the rocker arm oil line through the proper hole in the head. BE CAREFUL! Screw the rocker arm long studs into the head and place the rocker arms onto the head. If the head has been milled more than .060" we usually place spacers equal to the mill, or slightly less than the mill under each rocker arm.

Insert pushrods through head and into cups on valve tappets. Install rocker arm oil line into proper fitting. Back off all tappet adjusting screws to full loose position. Tighten rocker arm stand cap screws and the two nuts on the studs. We are now ready to set the valve clearances.

Valve clearances can be set easily by watching the tappets. Turn the engine over easily until the tappets for No. 6 cylinder are rocking, that is, the exhaust rocker will have just closed and the intake will just start to open. Then set the intake and exhaust valves on No. 1 cylinder. This procedure can be followed throughout the process of setting the tappets. After setting No. 1, watch for No. 2 to rock and set tappets on No. 5 cylinder. Rock on No. 4, set No. 3, rock on No. 1, set No. 6, rock on No. 5, set No. 2, rock on No. 3, set No. 4. There, it's finished! Remember to set your racing cam with the clearances supplied on the cam tag by the grinder. If you set them at stock, the engine will not develop full power, and will burn valves as they will not seat properly.

O.K., let's get this outfit on the road!

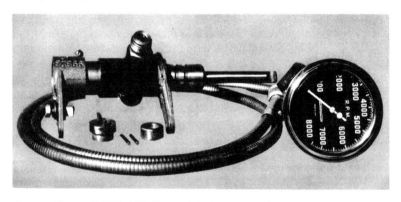

Stewart-Warner O-8000 RPM Hot Rod Tachometer, with drive and cable to fit Chevrolet or GMC engines. This tachometer is accurate at all speeds. A new S-W electric tachometer is out, which does not require any special drive equipment.

FIRING IT UP

1. Put oil in crankcase
2. Turn pump with extension on hand drill until pressure shows on gauge.
3. Make sure that there is water in the radiator; loosen temperature fitting at rear of head until some water appears. This eliminates any chance of an air lock which would later cause a steam pocket.
4. Set engine so that #1 cylinder is on compression stroke. If you have a flywheel on the engine (with an indicator) set the pointer on the ball.
5. Insert a strip of thin cellophane (such as that on a cigarette package) between the breaker points. Rotate magneto or distributor body until cellophane just slips free of points. Rotor should be pointing to #1 cylinder. Fasten distributor or magneto securely. If using a two-point distributor, insert cellophone between points which break last.
6. Wire ignition cap in a clockwise direction. Firing order: 1—5—3—6—2—4
7. Attach primary connections to coil, and hi-tension lead from coil to distributor
8. If using a magneto, omit Step #7 and attach grounding switch lead to terminal on side of magneto. No current of any kind should be put into magneto!
9. If engine is new, use two batteries in series (12 volts total) to provide starter current.

Special dash, hand made by Dan Stropel for his 1948 Chevy Sedan. Dash is engine turned, uses Stewart Warner instruments.

PLANNING

It is nearly always a good policy to try to build up your engine on another block, especially if you have but one car. Trying to build up a hot, or even "semi-hot" engine in a hurry is a mistake! Try to find yourself a 1941 or later block and crank assembly if you possibly can, as these are usually equipped with thicker cylinder walls than the earlier models. In order to aid you in your search for a block and crank assembly, the author has included a table showing the casting numbers of these blocks.

Hi-torque engines have a 3 - 9/16" bore with a 3 - 15/16" stroke. The stroke can be measured with a scale to check which assembly you are getting. Not only that, but all hi torque cranks have a raised "T" on the center counterweight of the crankshaft. This will aid you in finding a block and crank assembly. Used hi-torque short blocks are worth from $15 to $25 depending on their condition, and a good used 1941 or later short block (216" standard models) should not cost over $15.00. By short block we mean: crankshaft, block, cam, front mounting plate, pistons and rods. Prices quoted are for used assemblies with no holes in the block casting or cylinder walls. Most Chevrolet dealers will sell you a used short block for $15.00.

CHEVROLET BLOCK ASSEMBLIES

MODEL	TYPE	CASTING NUMBER ON BLOCK
1937-39	STD.	838941
1940	STD.	839132
1941	STD. 216"	839400
1941	TORQUE 235"	3660439
1942-47	STD. 216"	839770
		839910
1942-47	TORQUE 235"	3835335
1948-51	STD. 216"	3835353
1948-49	TORQUE 235"	3835309
1950-51	TORQUE 235"	3692703

CHEVROLET CYLINDER HEADS

MODEL	CASTING NUMBER
1937-40	838773
1941-48	839401
1949-51 std. 216"	3835409
1950-51 235" & P.G.	3835499

Equipping Your Chevy With A Racing Cam

Since the production of Chevrolet Manual II, the development of high performance camshafts for Chevrolets has continued. Now, cams are available which provide truly terrific performance.

It has come to our attention that a great many of our readers do not understand the functioning of a cam and for that reason we have attempted to include in this manual a complete explanation of the functioning of a cam, as well as showing by means of drawings, the timing comparisons of several camshafts.

The stock Chevrolet cam timing has been chosen by General Motors engineers as ideal for producing the maximum low speed torque from the Chevrolet stock carburetion, compression ratio, engine size and general passenger car and truck use. Quietness of operation with low spring tension and maximum economy have been given top consideration.

Naturally, if more speed is to be desired then it will be necessary to take into consideration the fact that we are going to change a few things that the stock engine was designed to do. However, in a Chevrolet, G.M.C. or Buick engine, the idling smoothness and low speed torque are not impaired as much as with a V-8 flat head type engine.

The stock Chevrolet camshaft is designed so that the peak h. p. is produced at approximately 3400 RPM's. As the duration of valve opening is increased, overlap is increased and the lift is changed, the point at which your engine produces its maximum horsepower will move to a higher engine speed range. This means that your engine will not idle quite as smoothly as a stocker, nor will it have the characteristic of going along at 5-10 miles per hour in high gear. This will certainly not deter those looking for real performance.

When a cam is installed with the proper additional items of speed equipment, such as carburetion, compression and perhaps additional displacement, you will find that you have greatly improved the acceleration, and the speed available in every gear.

A camshaft in a Chevrolet can be compared to a man who has just stuffed himself at a holiday meal and loosens his belt so as to be able to breathe easier. You have perhaps noticed how Chevys

G. M. C.'s and Buicks all seem to be quite "busy sounding" as they get up to high (for stock) speeds. With a cam, this all changes, and the top speed is not only improved, but the engine actually seems to be loafing at speeds where, with the stock cam, it seemed to be working very hard. The author's own coupe with a stock engine seemed to be really working at 60 MPH, whereas the engine now seems to be happiest at speeds approaching 85 MPH, even though it certainly runs quite smoothly at any speed from ten miles an hour on up in high gear, and the acceleration is phenomenal.

The basic function of a racing camshaft is to allow the cylinders to fill more completely, and by alteration of the opening, closing, and lift of the valves, the cam provides greater volumetric efficiency for the engine with only slight sacrifices in lugging ability and idling characteristics

The charts included with the dynamometer tests show that a racing cam with carburetion, and perhaps compression, will produce more horsepower than stock throughout the range of engine speeds even though the maximum h.p. is produced farther up the range.

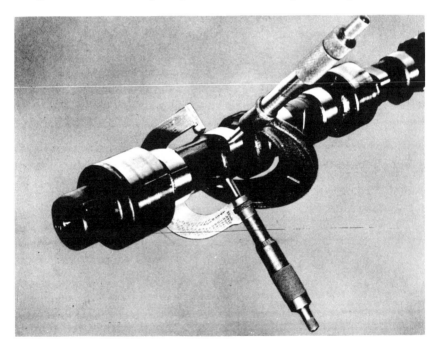

In order to figure the lift of a cam used on a Chevrolet or G.M.C. engine it is only necessary to "mike" the heel of the cam and subtract this figure from the dimension obtained by miking the lobe at maximum lift. Multiply remainder by the rocker arm ratio of 1.4 to get the approximate lift.

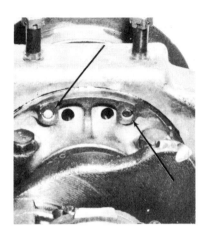

Arrows indicate holes to be tapped for 3/8-16 National Coarse Thread. This is a great aid to camshaft installation, especially where several cams are being tried.

With the #1 piston on compression stroke, at T.D.C., and the valve clearances set at the TIMING CLEARANCE recommended by the cam grinder, insert a .001" or .002" feeler gauge under the exhaust rocker arm. Turn the engine in the direction in which it runs. Keep turning until you can just slide the feeler gauge (some drag should be apparent). Note the timing on the pulley and mark it down, as so many degrees Before Bottom Dead Center. Keep turning the engine, leaving the feeler under the rocker. When the feeler is released and has a slight drag on it as before, again mark down the timing, which will be a few degrees past Top Dead Center.

Back up the engine until the gauge can be inserted under the intake rocker arm, and turn the engine slightly back and forth until you can notice a slight drag on the gauge when you are sliding it out. Note the timing, it will be Before Top Dead Center.

Leave the gauge under the rocker, and turn the engine over in the direction in which it runs, until the gauge is once again released with a slight drag. Note the timing, which will be After Bottom Dead Center.

This timing should correspond closely (within two degrees) to that of the cam grinder. If you don't obtain this timing, check through the whole procedure again very carefully.

If the intake valve, for instance, opens later than the cam grinder intended, then the cam is retarded. If the intake valve opens earlier than it is designed to do, then the cam is advanced.

Moving the cam gear one tooth advanced will advance the valve timing of the entire cam approximately 14 degrees. Moving it the other way will naturally retard it by the same amount.

While the above operations will take time, certainly it is important to know just what timing you have and whether or not it is "right on."

How To Select A Cam

The selection of a cam for your engine is important. If in doubt as to which will be best for your engine, its use, and the equipment you plan to install, contact the author for a recommendation. Each and every cam grinder will naturally feel that his cam is the best for your engine. We have tried nearly all of them. Our best results have been obtained with the cams listed in our catalog.

It is important to remember that once you have tasted the sweet fruits of more horsepower, that the craving will increase. For that reason, we usually advise a more radical cam than a ¾ for the fellow who is really interested in speed. Read the descriptions carefully. They may vary with some makers cams.

SEMI—Discontinued by most cam grinders for Chevrolets, GMC's and Buicks, as these engines idle well and perform best with at least a ¾ grind. The average semi cam will not provide enough improvement in horsepower to warrant the installation. If you want only a mild improvement in acceleration, then install a set of BARKER Hi-Lift Rocker Arms.

¾ GRIND—This is the most popular grind on the market for the three engines discussed in this book. When used with dual carburetion and added compression, these cams will provide a tremendous improvement in acceleration. Usually as quiet as a "stocker" the ¾ grind idles smoothly, accelerates well, and gives more top speed. Be sure to see the dynamometer test charts!

FULL RACE—Designed for dual or triple carburetion, oversize bores, and higher than stock compression. These cams will give more acceleration and top speed than the ¾ grinds.

SUPER—Many fellows run these cams in their street jobs. The author has been using a super grind in his own street job for almost two years. The engine is terrific for traffic and superb for high speed road work. The average super grind cam will idle rougher than a stock cam, but with overhead valves, this is not as apparent as with a flat head, V-8 type engine. Super cams should be used with big bore Chevrolets or Buicks, and large bore GMC's. Their longer duration and higher lift helps a lot to improve the breathing of a "big" engine at high speeds. The super grinds are sometimes not quite as quick on acceleration from a standstill if the engine is driven like a stock one. But, if the RPM's of the engine are kept up to 2000 or more until the clutch is fully engaged, the engine will have enough torque to really get your outfit underway. Note: Our dynamometer test engine produced as much horsepower at 3000 RPM's with 232" as the average, good running 3/8" x 3/8" (296") Mercury. This was with the HOWARD SU9A camshaft in the engine.

TRACK—Track grinds are designed with a long, flat power curve which gives a reasonably flat response within a given range of engine RPM's. This is usually over the 3500 to 4800 RPM range. Check the dynamometer charts for further graphic explanation of this characteristic of these cams. Usually these grinds are designed for maximum volumetric efficiency, and require high spring pressures. They are nearly always a noisy cam. Their use is not recommended for a passenger car!

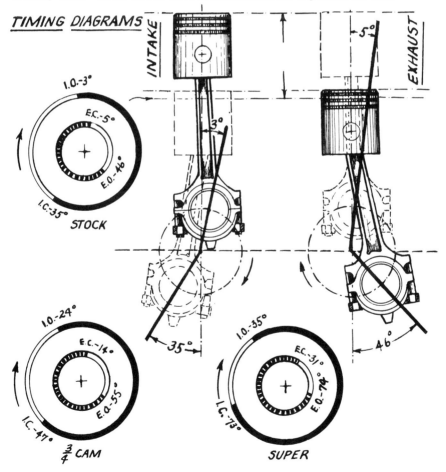

TIMING DIAGRAMS

S.C.O.T. supercharger installation. Made on a 1950 Chevrolet Power-Glide owned by Luigi Borelli, California Bill found this to be a difficult installation due to the tremendous amount of modification necessary and failure of Italian maker to include necessary items in the $467.00 kit. Carburetors are ZENITH 8810 side-draft type, with flexible hose to air intake. Pushrod and rocker-arm covers are by NICSON, ignition is MALLORY throughout. Head ported and polished by the author.

Cam Functioning

On the stock cam, the intake valve opens at approximately Top Dead Center (T.D.C.) and closes 35 degrees after B.D.C. As we want our engine to operate in the high speed range, it will be necessary to open the intake valve earlier and close it later, thus effecting a longer duration of valve opening.

Not only that, but we will increase the acceleration of the valve as it leaves its seat and as it returns to it. This acceleration is governed by the contour of the cam lobe at the points known as the opening and closing ramps. The acceleration of the valve off of the seat must be within the limitations of the valve operating gear, that is the pushrod, valve tappet, rocker arm, valve and springs. If the valves are opened too quickly, broken tappets will be the result, and too fast a closing action will cause broken rocker arms.

In order to keep wear and noise to a minimum, the racing cam should be equipped with opening and closing ramps which reduce the impact loading on the valve as it is raised from and returned to its seat, thus insuring reliability of the valve mechanism.

Now, back to the intake stroke of the piston. Shortly after the piston begins its travel down the cylinder, the pressure within the cylinder drops below that of the outside atmosphere, causing a suction within the engine (partial vacuum). At B.D.C., the pressure within the cylinder is still below that of the atmosphere and the inrushing column of gas has a certain amount of kinetic energy (momentum) which causes it to continue filling the cylinder with fuel even after the piston starts its upward travel. Thus, we hold the intake valve open past B.D.C. in a racing cam. However, this closing cannot be delayed too long or the gas will be blown back into the manifold and out of the carburetor. Delaying the closing past B.D.C. causes a racing cam to show less compression when checked with gauge.

Leaving the intake valve open past B.D.C. provides a moderate form of supercharging by allowing the inrushing gas and the partial vacuum within the cylinder to give more complete filling of the cylinder.

From the beginning of the compression stroke, the pressure within the cylinder rises to equal atmospheric pressure and then rapidly increases to 120-200 lbs. per sq. in., depending on the compression ratio employed within the engine.

Shortly before the end of the compression stroke, the spark ignites the fuel. If it were possible to burn the whole charge of fuel at once with no delay, this spark could be delayed until T.D.C. However, the flame takes time to travel through the charge of fuel within the combustion chamber, and this time increases (in relation to the fraction of time occupied by one stroke) as the speed of the engine increases. Thus, it is necessary to have more spark advance at top speed than at idling or slower RPM's. For that reason, most distributors and magnetos have a system for advancing the spark as engine RPM's increase.

As the charge is ignited in an engine of 7:1 compression ratio, the pressure increases to about 700 lbs. per sq. inch very rapidly. As the piston moves downward, this pressure decreases rapidly . . . to about 50-100 lbs. sq. in. at the time the exhaust valve opens before the end of the power stroke.

If it were possible to properly get the gas out of the engine, the proper time to open the valve would be at the end of the power stroke, as some power is lost by opening the exhaust valve early. However, this loss of power is not so great as the loss of power which would result from the braking action of the exhaust gas against the piston as it starts the upward exhaust stroke.

The exhaust valve is therefore opened before the beginning of the exhaust stroke of the piston and is held past T.D.C. over into the intake stroke period. This is known as "top overlap," and is built into a cam in order to get a scavenging action to clear the combustion chamber of burnt gases by the incoming flow of fuel mixture.

By leaving the exhaust closing until after the intake valve has opened, it is possible to fill the combustion chamber with fuel. If this were not done, exhaust gas would remain in the combustion chamber and mix with the incoming charge to weaken its potential power. The exhaust stroke of the piston can only clear the actual cylinder (or swept volume) of exhaust; the rest must be swept out by the scavenging action already mentioned.

For reasons mentioned above, most engines will run best with individual exhaust ports for each cylinder, each having its own individual exhaust pipe to aid in inducing suction in the cylinder for a complete scavenging of the exhaust gases. English motorcycle designers carry this to a fine point by actually tuning the length of the exhaust pipe to obtain an additional supercharging effect by placing a terrific suction on the cylinder with an exhaust pipe of the proper length.

One other point will be of interest to those readers who are engineering-minded. The intake valve is opened earlier for another reason besides those just mentioned. Intake valves must be opened so that they will be well off of their seats when the piston reaches T.D.C. and fully opened by the time the crankshaft is traveling at its maximum speed on the downward stroke. This insures a maximum filling of the cylinder with fuel.

Line boring Chevrolet main bearings. Note sturdy jig used to insure accuracy.

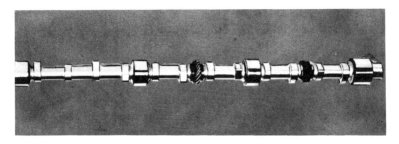

Relieved camshaft, Super grind. It is necessary to relieve these cams between the lobes so that the grinding wheel will not hit the cam bearing edges and bend the cam as it is ground.

HOW TO INSTALL A RACING CAM

When installing a racing cam in a Chevrolet, it is first necessary to gain access to the front of the engine. On all models 1937-41 this can be easily accomplished by removing the fenders and radiator as an assembly, rather than removing the radiator and grill alone. Access on 1942-48 models can be easily obtained by removing the grill and radiator assembly, as the fenders are difficult to remove. 1949-51 models present an additional problem, in that the center pivot arm of the steering must be removed before the pin can be dropped out of the way.

After the pulley is off, remove the timing cover. Turn the engine over until the two holes in the cam gear coincide with the two bolts or screws which hold the bronze cam retaining plate. Remove these screws. Remove the rocker arm assembly, pushrods and tappets. Make sure that the engine is on T.D.C. for #1 cylinder. Remove the camshaft from the engine.

In order to remove the crank pulley, a special puller is necessary. Do not attempt to remove the pulley at any point other than with the screw holes which are placed there for that specific purpose. We have tried other ways, and each time the pulley has been fouled up in some way before it finally came off.

The pan must be removed from any stock Chevrolet in order to get at the cam, due to the fact that two of the screws which retain the timing cover are located inside the front main cap.

Tap the holes in the front main cap for 3/8 - 16 thread and drill out the two corresponding threaded holes in the timing cover to clear 3/8" cap screws. These screws should be ¾ to 1" long.

Install a new crankshaft oil seal in the timing gear cover. Install aluminum timing gear on camshaft and steel gear on crank. Old crank gear can be removed with same puller used for the crank pulley, except on some older models without threaded holes in the crank gear.

For in-installation of cam, see data on "FIRING IT UP."

Valve Components

The various valve components of an overhead valve engine are far more important than those for a flat head. Since there are more basic parts, it is necessary to pay more attention to each one. For the basis of our discussion in this chapter, we shall be directly concerned with the valves, valve springs, various washers for the springs, spring retainers, keepers, rocker arms, rocker arm stands and shafts, pushrods, and valve tappets or lifters.

VALVES

Stock valves will be used by most of the readers of this manual. However, several modifications are possible where larger valves are desired. For instance, the 1950-270 GMC intake valves can be installed (with care) in the earlier, Group I cylinder heads. Power-Glide intake valves can be installed in 1937-40 Chevrolet heads, but the installation is not recommended for 1941 or later Chevrolet standard 216" heads.

Large exhaust valves can be used to advantage in the late 270 GMC heads. We suggest the use of the GMC #306 exhaust valves. The guides must be reamed out to make the installation, and different spring retainers and keepers will have to be substituted. We have used 1941 Cadillac valves which we have chucked in a lathe and turned down to the needed diameter. New keeper grooves are then cut in the proper position. When large exhaust valves are installed in late GMC heads, the insert valve seats must be removed, the port widened out, and the valve seat relocated.

Several builders have been using 1949-50 Chevrolet intake and exhaust valves in their Chevrolet heads, as this installation will serve to slightly increase compression. This is due to the fact that these valves are flat across their heads instead of concave. When using 1949-50 exhaust valves, special attention should be paid to the seat angle. If you are using an older head, the seat angle must be changed to 45 degrees, or the exhaust valves refaced to 30 degrees.

We prefer 270 GMC exhaust valves for use in Chevrolet heads. These have been really proven in severe competition and will usually last longer than comparable Chevrolet valves. These will fit Chevrolet heads with no alterations. Check seat angle of valve and head before completing installation.

VALVE SPRINGS

The choice of valve springs for a reworked Chevrolet head and engine is often a difficult one to make. It is important to get sufficient spring pressure in order to keep the valves from floating at high engine RPM's. If this tension is too great, it will demolish the lifters in a hurry, if it is too light, the valves will float and the lifters will again be allowed to go to pieces as they beat against the cam. The use of a really stiff spring combination will increase valve noise.

Sometimes we are asked just what valve float sounds like. In really severe cases it is quite noticeable, as it sounds like a hail-storm on a tin roof. In minor cases, floating valves cannot be detected unless the engine is placed on a dynamometer.

Several combinations of valve springs are available. Where an inner spring is used, it is very important that you make sure the outer springs are wound in the reverse direction of the inners. Chevrolet springs are available wound in two directions.

The mildest spring to use is the Chevrolet stock outer by itself. Apparently the Buick inner and outer have the same spring tension as a good Chevrolet outer, but we feel that the combination of two springs is better than the stock Chevrolet outer, as it seems to remove some of the harmonics apparent in a single spring. Next in order is the use of the Buick outer with special competition inner spring. Further tension can be had by using the Chevrolet outer with Buick inner. This combination must be used with care, however, as it will "bottom" when used with cams having a high lift. In order to check whether or not this set up will bottom, install the rockers on the head and the pushrods for #1 cylinder. Turn the engine until the intake valve is at maximum opening. Check the inner spring. There should be at least 1/32 to 1/16" between each coil of the inner spring. If not, then a different inner will have to be used. Our special competition inner springs are designed so that they can be used with maximum lift and not become coil bound or "bottom." These special inners with Chevrolet outers will give maximum valve spring tension. The set up is ideal for a competition engine.

SPRING HEIGHT

Spring height is very important. With the intake or exhaust valve on the seat, the height of the spring from top of the spring to the bottom should be 1—27/32". If it is more than this, then special tnesion washers must be installed to bring the spring to the required height. If it is less than this, then the seat can be lowered into the head, or washers can be removed from under the spring. For really wild cams and high RPM's, this height can be reduced to 1—25/32".

SPRING RETAINERS

Special competition retainers or spring seats should be used for racing engines. These can be used on the intakes and exhausts on GMC heads. They can be used on intakes only with the Chevrolet head. We suggest the use of 1949-51 valve spring retainers on the exhausts. Buicks have good stock spring retainers. Special ones are not necessary for the Buicks.

The reason that these special retainers are used for Chevys and GMC's is that the stock type are of pressed steel, and will let go at high speeds. Then the valve pulls the keepers through the retainer, and you have a cylinder demolished in the process.

ROCKER ARMS

Stock rocker arms must be used for Chevrolets and Buicks. Hi-Lift type rockers are available for the Chevrolets. These can be installed on a stock engine, or in some cases will provide additional performance where a racing cam is already installed. These intake rockers will fit GMC's.

Your racing GMC engine should be equipped with the new sheet metal type rocker arms. These are far stronger than the cast type and will stand up better for competition use. Obviously, the cast type were too weak, or the GMC factory would not have changed over.

ROCKER ARM STANDS

Special attention must be paid to the rocker arm stands when the head has been milled. We suggest that if your head has been milled more than .062", that washers be placed under each rocker stand to make up for the milling. This allows the stock height push rods to be used. Washers used under rocker stands must be exactly the same under each stand to keep the rocker arm shaft in line.

PUSHRODS

Special tubular pushrods are available for Chevrolets and GMC's. These should be installed, as they are 30% lighter than stock pushrods, and will not bend, even under severe spring pressures. Buick stock pushrods are of adequate strength for racing installations. They should not be used with Chevys and GMC's, as the cups at the tops of the Buick pushrods do not fit the Chevrolet and GMC tappet adjusting screws.

TAPPETS

Tappets (cam followers or lifters) are the biggest trouble with hopped up engines. The stock ones are all that are available at the present time.

For the last lakes season we had a special set of lifters made with CARBALOY tips. These did not wear well, as they were heavy, required excessive spring tension, and wore the cam lobes to an alarming degree. We have since discontinued their use, since their qualities did not warrant the price of five dollars each.

Our best luck with tappets has been had by using old tappets which have been run in engines for several years. These are coated with a glaze of dirt, grease and carbon, and will outwear five sets of new tappets. We prefer the 1937-40 style lifters due to the fact that these are lighter than the later models. These have the cup hollowed out within the tappet, and the pushrod extends into the tappet more than an inch. The later type uses a shorter pushrod and the cup for the end of the pushrod is on top of the tappet.

We suggest that you give careful attention to your tappets and examine them often. If you find that they are not giving trouble by breaking or galling, then just leave them alone, and chances are that they will never cause you any trouble. Above all, don't install a new set of tappets in your engine. It just isn't good for the engine, as they will wear out very quickly, and perhaps cause you to lose a race.

ROCKER ARM COVERS

Cast aluminum, polished rocker arm covers are available for GMC and Chevrolet engines. None are available for Buicks. Some have managed to quiet their valve mechanism by coating the rocker arm cover with underseal material. Since oil or gasoline will dissolve this material, installing a rocker arm cover of heavy aluminum is the best plan, since it does provide a better quieting as well as insuring that you will have a clean engine.

If you decide to chrome your own rocker cover, then it is not advisable to weld shut the vents in the cover and try to get a smooth looking cover. Vents in a rocker cover are a necessity, as they provide an escape for moisture which would otherwise serve to rust the rocker arm shafts and rockers in a short time if no vent were provided.

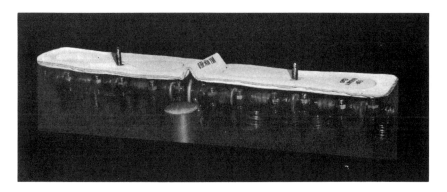

Top is the Val-Vin-Hed Silencer for Chevrolet, Buick and GMC. This pad eliminates up to 50% of valve noise and provides better lubrication. Middle photo shows Chevrolet engine equipped with NICSON alumnium rocker arm cover. This ⅜" thick aluminum casting provides the utmost in valve silencing. Lower photo is of NICSON aluminum pushrod cover. Also available for Power-Glide and GMC engines.

Rear view shows sweeping body lines provided by the Bantam roadster body. Wheel wells have been filled in neatly by contoured metal panels which carry out the body lines perfectly. This car receives favorable comment wherever it is shown.

The front view of Healey's roadster clearly illustrates the low center of gravity and comfortable seating position of this car. Note tubular shock absorbers and semi-elliptic suspension of front axle. Radiator shell is from 1932 Ford.

1933 Chevrolet engine used by Mr. Meb Healey in the roadster described in the accompanying article.

MEB HEALEY'S CHEVROLET ROADSTER

This car has attracted much interest whenever it has been run. The body was constructed from a BANTAM roadster body, with the fender wells attractively faired-in and the hood and grill custom made by Mr. Healey.

One of the first fast Chevrolets in Southern California, this car features a 1933 Chevrolet engine. Mr. Healey feels that the long stroke of the engine and its small displacement have reached the limit of development and is now planning to install a later Chevrolet engine for the coming Lakes season.

The manifold was hand made, as was the special, all ball-bearing two point two coil distributor. Two D.S.M. coils were used with D.S.M. condensers.

The car was never run on alcohol fuel at the Lakes, so its potential top speed will never be known. A time of 110 mph has been recorded on gasoline. The acceleration through the gears is really breath-taking. The top speed performance is silky smooth. The engine is coupled to an overdrive unit giving a very high final drive ratio.

We feel that this is a true American sports car, as it will start and run smoothly on gasoline and yet has a top speed of over the century mark, coupled with fine handling characteristics. This car could be a threat to the Allards and Jaguars if ever entered in road racing events with a hotter engine.

Mr. Healey's factory has absorbed all of his time for the past two years and the car has not been run during that time. However, it is not for sale, as he definitely plans to make improvements and use the car again for some really exciting motoring!

Pistons & Piston Rings

When you stop to consider the amount of force which each piston in an engine is called upon to bear, it becomes quite obvious that the pistons for your hopped up Chevrolet or GMC engine should be chosen wisely.

For normal road use, with occasional drag racing and top speed work, the aluminum split-skirt T-Slot type replacement pistons will serve the purpose well, and will also provide adequate oil control, as these pistons may be set up with rather close clearances.

Where all out racing is to be the primary use for your engine, then oil control does not loom as a large object, and solid skirt pistons should be used. Solid skirt pistons must be fit with greater than stock clearances, as they don't have slots in them to compensate for the expansion of the piston as it heats up in the engine.

For track use, marine racing, and straightaway, the T-Slot piston need not even be considered, as it will be likely to let you down in a moment of extreme stress.

Cast-iron pistons may be used in a street or road job where high speeds in the gears are not contemplated and where the driver is very cautious about his engine. What most of our readers do is to install T-Slot aluminum pistons in place of the cast-iron pistons, due to the fact that the aluminum pistons are easier on the rod and main bearings since they are lighter than the cast iron stock models.

The cast-iron pistons have been known to come apart at the point where the lower ring groove is cut. We do not recommend their usage where you want your engine to really perform, as the breakage of a piston can be quite serious.

Clearances on the skirt of the piston will vary with the type of use, as is shown in the accompanying table. Close attention to these will provide longer trouble-free life for your engine.

Ring groove depths are another important part of fitting pistons and rings in your hot engine. When the compression rings (the top two rings of each piston) are inserted in the groove as shown in the photographs, the edge of the ring should come flush with the edge of the ring groove or no more than .010" below it. This should be

Compression ring inserted in ring groove which is too deep. Note that edge of ring comes far below the edge of the groove. No expander should be placed behind the ring to make this check. Shims will be installed to decrease ring groove depth.

Two shims were needed to bring the ring (shown in the first photo) to just slightly under the edge of the groove. These shims are shaped by hand, and then laid in the groove of the piston. Only the top two grooves of the piston receive this treatment. No shims should be used in oil ring grooves.

Here one shim has been laid in the top ring groove. Note that there is clearance between the ends of the shim of $3/8''$. These gaps should not coincide where several shims are used in one ring groove.

Now the piston ring fits the groove properly as regards depth. Some rings are available to fit oversize pistons with deep grooves. Grant refers to these as "Deeps" as they will fit a deeper ring groove adequately.

measured with the bare ring in the groove. No spring or expander should be behind the ring when this check is made. Most replacement pistons will have ring grooves which are quite a bit too deep. Where this situation is encountered, it is necessary to use ring groove shims. This shim stock is sold by the pound and is made especially for the purpose.

When shim stock must be used to lessen the depth of the ring grooves, then you should carefully measure the amount of shim stock necessary to fit closely in the ring groove and still leave about ¼" maximum clearance between the ends of the shim with the shim wrapped around the piston. Careful shaping of the shim with your fingers will allow the shim to lie in the groove, hugging the inside of the groove closely. UNDER NO CIRCUMSTANCES SHOULD THE SHIM STOCK BE WOUND AROUND AND AROUND IN THE GROOVE. Individual shims must be cut for each thickness of shim stock which is necessary. To determine how many shim thicknesses are necessary, check the ring in the groove for depth as was mentioned previously. Then cut two or three short lengths of shim stock about 4" long, and place these under the ring one by one until you determine the number necessary to make the groove the proper depth.

Robert Powell's VICTRESS Sports Car holds Santa Ana Drag Strip Record for top qualifying time of over 111 m.p.h. Dry Lakes runs show top speed to be in excess of 150 m.p.h. on gasoline. Engine is G.M.C., equipped with FISHER head and pistons, SPALDING cam, FORD coil, and California Bill's special distributor. Fiberglass body by Boyce-Smith.

Using feeler gauge to check ring gap width. Ring must be squarely in the bore to make this check.

Prior to using this procedure, most large bore Chevrolet engines would turn into oil eaters in a very short time after the engine was built. Since using this type of set up for ring grooves and piston rings, no more difficulty has been experienced with excessive oiling.

Ring gap must be carefully checked, even when the rings are supplied to fit a specified bore. Grant Rings specify a ring gap of .003" per inch of cylinder diameter. Thus, a cylinder such as the stock Chevrolet Standard engine would require a ring gap of 3.5" x .003" or .012" when figured to the closest thousandth. This is a minimum figure and should not be lessened, due to the danger of sticking a ring in the bore, which can really mess up a cylinder wall in a hurry.

Other factors which enter into the fitting of piston rings are usually not considered, but they are important nevertheless!

The SIDE CLEARANCE of the ring in the groove should not be less than .0015", nor should it exceed .006". This latter figure of .006" should not be considered a recommendation as to setup, but is a maximum wear figure which, if exceeded, would necessitate replacement of the piston or use of a wider ring.

When replacing rings, check for a ridge in the ring groove in the old piston. If a ridge or ridges are present, either remove them or install new pistons.

It is exceedingly important to check the oil ring groove to make sure that the return holes or slots are clean and free from carbon deposits before installing the rings on the piston.

Marvin Lee's GMC streamliner was wrecked on its maiden appearance at the Bonneville Speed Trials in 1950. The car was clocked at 197 mph on its warm up run and was reputedly travelling approximately 216 mph when it flipped. Luckily, Puffy Puffer, the driver, escaped with only slight scratches.

Special Heads

The stock cylinder head on a reworked full-race block assembly will produce .8 to .91 h.p./cubic inch when set up for nitrated racing fuel. Whether or not you will be better off financially by using a stock or special head depends on the horsepower you want, and how far you intend to go toward making the engine perform. Special heads for the Chevrolet and G.M.C. engines start to perform where the stock head quits. For example, the stock 270H cylinder head, modified for larger valves and using more compression, when used with a completely equipped 292 cubic inch racing lower end can produce 285 h.p. on nitro-methane and methanol if super-tuned. By way of contrast, an aluminum head such as the FISHER 12 PORT GMC will let the almost identical engine produce 292 h.p. per cubic inch, before the nitro-methane is added!

47

274 cubic inch GMC engine built by WAYNE F. HORNING, designer of the FISHER 12 PORT GMC HEAD. Designed for the 1950 Indianapolis race, this engine was installed in a car too heavy to make the program, even though output was 285 h.p.

It has been the author's personal observation that anyone purchasing one of the 12 Port heads (such as the FISHER, HOWARD, or WAYNE) should have a good store of basic tuning knowledge before attempting to build an engine with one of these heads. However, for any one wanting maximum performance from gasoline or special fuel, one of these heads and its accessory equipment will often prove to be a better investment when horsepower output is weighed against dollar cost.

If you are not limited in cubic inches, consider G.M.C. performance, with or without a special head, before spending a lot of money on a Chevrolet engine. No matter how many dollars you might spend, it is impossible to compete with displacement. More cubic inches are the cheapest way to additional reliable horsepower. This fact has been proven time and again.

If you already have a well-equipped Chevy engine, you can eliminate the necessity of having to dispose of much of your equipment (since the 12-Ports require special manifolding, exhaust, etc.) by purchasing one of the new NICSON aluminum heads pictured on page 54.

How To Rework Your Chevy Head

Reworking your Chevrolet head will fall into one of three classes, depending on the type of engine you plan to have when you finish.

The first, and most simple modification, consists of milling the head. Your cylinder head can be milled by almost any machine shop. Current prices for the operation are usually under ten dollars, even for a .187" mill. A chart has been prepared so that you can ascertain just what a given amount of milling will do for your particular engine.

The next modification which will be undertaken by many of our readers will be porting the head so that the intake ports and intake valve pockets are enlarged to provide for an unrestricted entry for the fuel mixture. This adds acceleration and greatly helps the horsepower output at high speeds by aiding the high speed breathing of the engine.

High compression pistons are now available, to eliminate the previous need for filling in the combustion chambers by welding. Thus, higher compression is easily obtained and requires only the installation of a set of new pistons. Two types are available, the 7.75:1 ratio split-skirt (T-Slot) for all road jobs, and the 8.25:1 ratio models for all-out racing. Various bore sizes are available.

Now we will attempt to break down each process into simple directions so that you can rework your cylinder head for the high speed performance that you are seeking.

MILLING

Remove the head from the car or engine. Remove the valves and springs. Wash it thoroughly with gasoline, solvent or a GUNK solution. Take the head to your local machine shop and specify the amount of mill desired. 1941 and later Chevrolet heads must have the intake valves sunk the same amount that is removed from the head, especially if racing cams are to be used. 1937-40 heads must not be milled any at all unless you are willing to replace the domed pistons with flat top pistons. Chevrolet flat top pistons, as well as the split and solid-skirt aluminum pistons will fit these earlier blocks with no modifications. If a 1937-40 head is milled as much .060" the domed pistons will usually hit the head.

COMPRESSION RATIO TABLE

HEAD TYPE	MILL	216"	235"	VOLUME
1941-51 Stock	.000 .0625 .125	6.5 6.9 7.25	7.0 7.45 7.75	6.54 6.08 5.78
'37-40 Stock	.000 .125 .187	6.0 7.15 7.8	6.46 7.7 8.4	7.18 5.84 5.31

Calif. Bill points out the "knobby" new McGurk Hi-Power pistons. Engine shown here was built by Calif. Bill for South American customer for installation in cross country road-racing coupe. Note fittings for full-flo filter, ported and milled head.

1937-40 heads can be milled .187" (3/16") without need for resinking the intake valves. These heads can be used on all engines 1937-51, except 1950-51 Hi-torque and Power Glide engines. 1937-40 heads must be equipped with 1937-40 intake valves.

Consult the accompanying chart of compression ratios to find what compression you can obtain on your engine by milling. If you have an oversized bore, your compression can be easily figured by following the data which we have included in "HOW TO COMPUTE COMPRESSION RATIOS."

PORTING

Many of our acquaintances owning Chevrolets have frankly admitted to us that they are afraid to try to port their heads for fear that they will go into water (break through the water jacket around the ports). By closely following the information which we have presented here, your chances of ruining your head are exceedingly slim. The walls of the intake ports are quite thick, as are those of the exhaust ports. Enlargement of these ports can be undertaken without danger of weakening the structural strength of the cylinder head.

The easiest way to port your head is by use of "shell reamers." These reamers are available at any machine tool supply house, although you may have to manufacture an arbor to fit your ½" electric hand drill. Usually, these same stores will have suitable arbors, which may need to be chucked in a lathe and the diameter of the shank reduced to fit your equipment.

When enlarging the intake ports, both in the side of the head and the valve pockets, do not use any cutting oil, and be very careful to get a good hold on the drill motor so that it won't "throw you" when the shell reamers catch in the cast-iron. If you don't do this you may get a nasty sprain in your arms and wrists.

The intake ports in the side of the head slope downward to the intake valves. It is advisable to follow the approximate contour of the original port when using the reamers, or even to point the reamer so that it cuts more out of the bottom of the port than out of the roof of the port. Don't get mixed up by terminology; just remember how the head will be installed on the engine and you'll see immediately just what is meant.

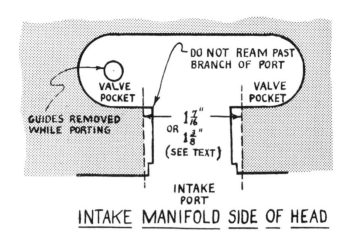

INTAKE MANIFOLD SIDE OF HEAD

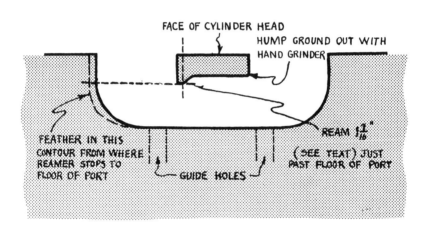

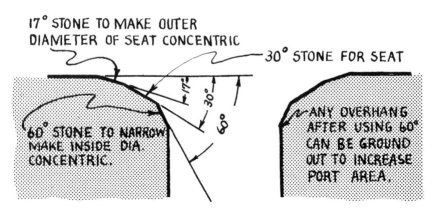

We recommend that you use 1-7/16" shell reamers on the intake ports and valve pockets on any head 1937-40. Use 1⅜" reamers on 1941-51 heads.

Ream into the side ports to the point where the branch begins. By branch we mean the point at which the port ends as a circular tunnel and branches to carry the gas to each valve. This will leave a ridge in the port, which should be feathered in with a hand grinder or flexible shaft tool.

Ream into the intake valve pockets approximately ¾", or to the point where you just clear the floor of the port. Here again you will have to feather in the rough edges with a grinding tool. It is a good idea to remove the valve guides before trying to ream into the valve pockets.

Exhaust valve pockets should be reamed out with a 1¼" reamer. Ream in to the point where the reamer goes just past the bottom edge of the port. Long experimentation has shown the author and all others associated with Chevrolets, that not much attention need be given the exhaust ports. Your best move as regards getting rid of the exhaust will be to pay particular attention to getting a good header system for the engine. Tests have shown that large exhaust and intake valves in a Chevrolet head will not provide any more horsepower at the top end, but large intake valves will improve the acceleration through the gears.

POWERGLIDE AND 1950-51 HI-TORQUE HEADS

When porting either of these heads (they are identical) you can use 1½" shell reamers into the side ports, and 1⅝" reamers into the intake valve pockets. Exhaust valve pockets should be given the same treatment as the earlier heads, using a 1¼" reamer. Rough edges should again be feathered with a hand grinder.

FINISHING

Finishing your porting job will take time. We usually estimate eight to sixteen hours for an inexperienced worker porting a Chevrolet head, especially when done carefully. Various shaped stones will have to be purchased for your hand grinder so that you can give a smooth finish to all contours and surfaces within the intake ports and pockets. It is suggested that you only knock off the rough edges in the exhaust ports.

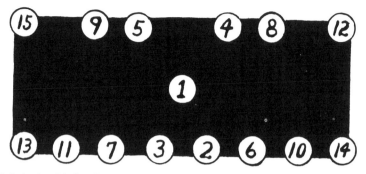

Cylinder head bolt tightening sequence. We recommend that all bolts be tightened to fifty pounds, and then taken down an additional ten pounds at a time in sequence, until all are torqued to 90 ft. lbs. If possible, torque on the head one day and check it the next day before installing rocker arm assembly. Otherwise check the head after the engine has been thoroughly warmed up.

The guides should be reinstalled after polishing the ports. Do not saw off the guides flush with the roof of the port! This will only cause the guides to wear out within a very short time. Actual tests have proven beyond reasonable doubt that sawed off guides will not produce any more horsepower.

After installing the guides the seats can be ground and the valves lapped in and installed.

To compute your final compression ratio, measure the volume of one combustion chamber and then consult the page giving the information on computing compression ratios.

The new NICSON aluminum head for all 1937-52 Chevrolets (except Power-Glide and 105 h.p. hi-torques) is equipped with insert valve seats cast into the head when it is poured at the foundry. Valves are stock Power-Glide, and design permits installation on the stock 3½" bore where desired. Weight is approximately one-third of the stock 70 lb. cylinder head. Any compression ratio 7.5:1 or higher is available. Head uses stock manifolding and rocker arms, so that it can be added at any time without scrapping equipment already purchased. Port area in this head is greater than any other head available for the Chevrolet block.

How To Compute Compression Ratios

1. Install valves in head.
2. Level head.
3. Fill 100 c.c. graduate with thin oil.
4. Pour oil into combustion chamber. Note amount left in graduate. Subtract this from 100 c.c. to get volume of combustion chamber in cubic centimeters.
5. Divide volume of combustion chamber in c.c. by 16.4 to get volume in cubic inches.
6. Compute total displacement of engine. See chart below.
7. Divide total displacement by 6 to get displacement of one cylinder.
8. Use formula below to figure compression ratio.

V_1 = Combustion chamber volume
V_2 = Volume of head gasket = 0.65 cubic inches
V_3 = Volume of one cylinder

$$\text{Compression Ratio} = \frac{V_1 + V_2 + V_3}{V_1 + V_2}$$

SAMPLE:

248" hi-torque (.090" oversize), volume of combustion chamber is 73 c.c. ÷ 16.4 = 4.45 cubic inches for V_1.

$$\frac{4.45 + .65 + 41.3}{4.45 + .65} = 9.1 \text{ to } 1 \text{ COMPRESSION RATIO}$$

CHEVROLET ENGINE DISPLACEMENTS

Bore	Std. 216"	235.5 Hi Torque
.060"	225"	244.0"
.090"	228"	248"
.125"	232"	252"

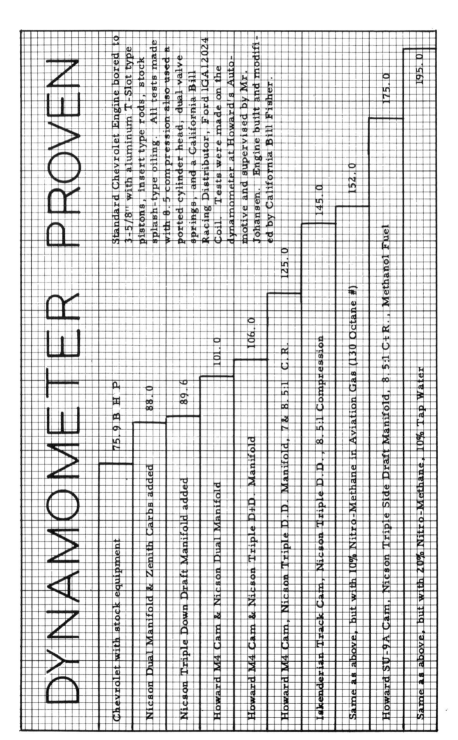

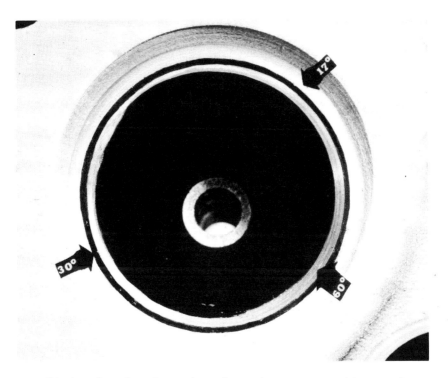

This photo shows the various angles used to produce a proper seat job on a racing Chevrolet or GMC head. 15 degree stones can be used instead of 17 degree if so desired.

SEAT DETAILS

In order to insure that your head will have maximum port area for good volumetric efficiency and excellent breathing at high speeds, it is imperative that special attention be paid to the width of the intake seats and the position of the seats in relation to the edge of the valve. To a certain extent this also holds true with the exhausts, but not to the same degree.

The best plan for getting the seats out to the edge of the valve (without a lot of facing, grinding with compound and making corrections little by little) is to notch a valve as shown in the accompanying photos. Then insert the valve in the guide of the port on which you are working. Check to see where the seat is located on the valve. If it is not far enough toward the edge, and it seldom will be, run the seat out farther with a 30 degree stone. Check again. When the seat is out to the edge of the valve, run a 17 degree stone on the outer edge of the seat to true the seat and keep it concentric with the edge of the valve.

Use a 60 degree stone on the inside edge of the seat to get this diameter concentric with the outer edge of the seat. The 60 degree stone can be used to narrow the seat after it has been moved out far enough on the edge of the valve. If any taper is left "hanging" in the port, this should be removed with a flexible shaft or hand type grinder as is used for porting.

This same routine can be followed for the exhausts, but it is usually not considered necessary. Width of the intake valve seats should be 1/16" to 3/32". Width of the exhaust valve seats should be approximately 1/8".

Notched valve used to indicate location of seat in relation to edge of valve. This eliminates seating the valve several times to get seat out far enough.

Ignition

Approximately one-half of the difficulties which arise in tuning a re-worked stock Chevrolet engine are a direct cause of ignition deficiencies.

Even the installation of a dual carburetor manifold will often show up ignition difficulties which were not apparent with the single carburetor.

Spark plugs are usually the first item which should be placed under suspicion. An approximate choice can be made from the table which we have included. When tuning a hopped up Chevrolet it is always advisable to use spark plugs at least one step colder than recommended by the factory or the spark plug manufacturer's specification chart.

If your engine tends to "run-on" after the switch has been turned off, it is fairly obvious that you need a colder spark plug, that your present spark plugs are completely worn out, or that you have some sharp edges in the combustion chamber which are getting so hot they act as glow plugs. Or, you may have excessive deposits of carbon in your cylinder head.

14 mm. spark plugs should be used in a hopped up Chevrolet head. If you have a 1941-1948 cylinder head, drill the plug holes out with a ½" drill and tap with a 14 mm. tap. You'll be glad you did! There are few heat ranges in the 10 mm. plug size and the small size of the plug tends to give very short life in a high speed, high compression engine.

The stock Chevrolet distributor is a good one. Its performance can be improved by reducing the point clearance to .012" instead of the recommended .018" indicated in the factory specifications. A larger condenser can also be installed on the outside of the stock distributor. The MALLORY is best suited for this as it has a large mounting ear which can be mounted under the clip used for holding the spring cap-clip. The stock Chevrolet coil will throw a good hot spark which is usually ample for a road job. For a hotter spark with less flashing at the points, the MALLORY "BEST" coil should be installed.

Ignition is one of the points often overlooked by the novice in tuning a high speed engine. The Chevrolet engine at high speed is very critical as regards ignition. The proper amount of advance and the settings to be used must be discovered and then used carefully from that time on if continued maximum performance is desired.

The MALLORY Co. has brought out a two point replacement kit which fits right into the stock Chevrolet Distributor and adapts the distributor to two point operation. In reality, this is the same principle which has been used for many years to give the coil more time for saturation so as to produce a hotter spark. One point is opened ahead of the other one by 8 degrees. One point is used solely for opening the circuit and the other for closing it. This greatly reduces point pitting and burning.

We feel that each and every Chevrolet on the road today can effectively use one of these MALLORY kits in its distributor, whether the engine is totally stock or completely hopped up.

Where the compression is increased to any appreciable degree, the MALLORY Two Point Conversion Kit can be installed to advantage. This kit will fit 1937 or later Chevrolet DELCO distributors and will also fit the distributors used on the G.M.C. truck engines. Tests show that the stock Chevrolet distributor modified by use of one of these kits will effectively fire 9:1 compression at 6000 RPM when used with a BEST coil, ECHLIN coil (with current ballast in top of coil), or a FORD 1GA 12024 coil. A current ballast should be used with either of the latter two coils to insure long point life.

The MALLORY distributor is equipped with the same two point set up as is used in the conversion kit. The distributor also uses a larger cam than the stock Delco-Remy part. Various ranges of advance are available in accordance with the type of engine and model which you are using. A fine feature of the MALLORY distributor is that the unit is sealed against moisture by means of a gasket in the two piece cap, as well as another gasket between the cap and the distributor body. These distributors work well on compression ratios to approximately 10:1.

The MALLORY BEST coil should be used with the MALLORY Conversion Kit or with the MALLORY distributor. For high speed and high compression, a BEST-X coil has been developed. The "X" denotes an extra 5000 turns on the secondary. Tests supervised by us indicate that the MALLORY BEST and BEST-X coils produce as hot a spark as any other premium coil on the market today and at the same time give less flashing at the points, an item to remember where long point life is desired.

Chevrolet Distributor Data, from Delco Remy Bulletin No. 1 D-3

1. Torsional vibration caused by loose or worn timing gears or a worn oil pump will cause rapid wear of the advance mechanism. To obtain normal life from the distributor, these conditions must be corrected.
2. Distributor oil pumping is usually caused by a crankcase pressure resulting from a clogged engine breather pipe. This can usually be remedied by cleaning out the breather.
3. Chevrolet type distributors are designed so that the end thrust is taken on the lower end of the distributor gear. If not installed with the distributor shank pushed down solidly into the mounting well, the thrust will be taken inside the distributor bowl and will cause the shaft to freeze!
4. The mounting must be clean so that the distributor will be properly grounded.

For all-out racing, the Spalding Brothers have developed a dual point distributor which is quite popular for severe racing use. They recommend its use with a pair of BOSCH "Big Brute" coils. The comparatively new MALLORY Magspark Transformer has also proved its worth for all-out racing when it is used with the special Mallory distributor or two point conversion plate.

Each builder usually has his own ideas as to the amount and rate of advance which should be used with his particular engine. Our 1941 coupe, with 248 inch displacement, super cam, 9:1 compression and dual carburetion works best when used with a 1937 DELCO Chevrolet distributor. This distributor has been modified with a MALLORY two point kit and has a total of 48° (forty-eight degrees) advance, or 24 degrees in the distributor itself. The vacuum advance is also used.

Other Chevrolet distributors have the following advance characteristics as shown in this chart.

MODEL	ENG. ADVANCE AT RPM	VACUUM DATA
1937-39	1.75° at 600 RPM	19° at 9" HG
649G or 1110008	50° max. at 3600	5" Hg to start
1940	4° at 800 RPM	16° at 12" Hg
#1110052	37° at 3200	6" Hg to start
1941-49	1° at 600 RPM	20° at 16" Hg
1110090	13° at 1200	7" to start travel
1112353	20° at 2000	
	39.5° at 3450	

As compression ratios are increased, the rate of advance should usually be slowed down accordingly. In order to reduce the rate of advance in a Stock Chevrolet DELCO distributor, it is necessary to lessen the weight of the counterweights by grinding them off, and a comparison of a stock and a reworked weight has been shown here. In order to insure that both weights are equal, clamp them together when grinding off the metal which you wish to remove. It is also advisable to weigh them after grinding to make sure that they are of equal weight. Otherwise, severe wear of the distributor bearings can occur if the weights are not equal.

Grinding the weights will not reduce the amount of advance in your distributor. If you feel that a reduction in the amount or total number of degrees will aid your engine, it is best to test first by disconnecting the vacuum advance mechanism and making several runs with the advance mechanism in the distributor. If an improvement is evident, try retarding the distributor and using the vacuum advance. If this reduces the performance, then try reducing the travel of the vacuum advance mechanism. This can be done by drilling a hole, or several holes in the arm which actuates the distributor, and inserting a large cotter pin so that the travel is restricted.

Above all, don't be afraid to experiment with the ignition. A few degrees advance or retard from the point at which your engine will run best will cause a tremendous power loss. A chassis or engine dynamometer can be a great help in getting your timing correct.

If you are using your engine for a great amount of racing and plan to tear it down often, a degreed crankshaft pulley should be installed and a careful note made of the point at which your engine runs the best as regards ignition timing. If you don't wish to buy a pulley which has been professionally laid out and degrees plainly marked thereon, you can fit a pointer to your timing cover and mark the pulley carefully with a cold chisel or center punch so that Top Dead Center can be readily identified.

Chopped and stock distributor weights from Chevy or GMC distributor.

SPECIAL DUAL POINT IGNITION

A good dual point dual coil ignition system can be built with very little expense if the enthusiast is willing to buy a Chevrolet type STEWART WARNER tachometer drive for approximately $18.00. The NASH dual ignition distributor (6 cylinder) that has an arrow on the distributor cam showing clockwise rotation is the one to buy from your local wrecking yard.

The collar should be removed from the drive end of the NASH distributor and the distributor shaft also removed. The case and shaft should be taken to a rebuilder and the old bronze bushings removed. New bushings should be installed and honed to obtain a perfect fit.

Install the shaft in the distributor, and place the dog driving piece (from the tachometer kit) on the end of the shaft. Install the new Chevrolet distributor gear on the end of the tachometer drive. Notice that the dog on the end of the distributor does not quite engage fully in the tachometer drive. This means that the distributor must have some metal removed from the case in order to fit into the tachometer drive as it should and also enable the clamp which holds the distributor to be used.

Now take the distributor cam and grind off every other lobe, being careful not to hit the lobes which will be used. AMERICAN BOSCH "BIG-BRUTE" or MALLORY condensers should be installed on the outside of the case, and the leads led into the points. With the right hand coil high tension wire in R and the left hand in L with the primary terminals on the correct leads, check to see if the rotor is in line with the 1R terminal inside the distributor cap. This is the end of the rotor without the brush. The points should be just beginning to open. Depending on which lobe of the distributor cam you started to grind with will determine whether the right hand points start to break on 1R or the left hand points start to break with 1L (brush end of rotor). Thus there are two possible firing orders for the distributor, depending on which point breaks (1L or 1R)

<pre>
     1R equals 1R-5L-3R-6L-2R-4L
     1L equals 1L-5R-3L-6R-2L-4R
</pre>

The NASH distributor has 12 degrees of centrifugal advance, which provides 24 degrees of flywheel advance.

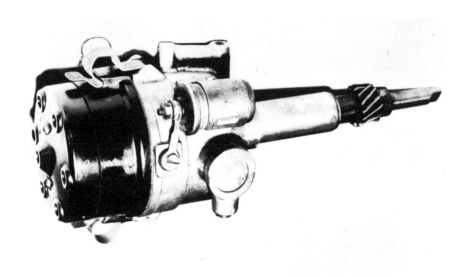

MALLORY distributor for Chevrolet or GMC. Available for many cars, these fine ignition systems feature dual points, balanced rotor, and water-resistant cap. Although excellent high-speed performance can be had by using these distributors with the stock coil, the MALLORY "Best" Coil or MAGSPARK Transformer is recommended.

Nash ignition converted for dual coil, dual point operation as outlined in the accompanying article.

Caution! Blasting Read With Care

The Chevrolet oiling system is an excellent one. If such were not the case, its use would have been abandoned long ago. A recent survey indicated that the Chevrolet engines used in fleet trucking service have had less rod bearing trouble than any other make, including those using full pressure oiling to the rod bearings.

Several years ago, when the author built his first Chevrolet engine for his street coupe, nearly all of his acquaintances in the field of speed made solemn predictions that the engine would blow up quickly if the stock splash oiling system were to be used. Having just been married, and with a lot of expenses to meet, the prospect of a blown up engine did not make the outlook very bright. However, one or two older and wiser men indicated that the stock oil system would work out fine if set up properly. AND—IT DID, AND HAS EVER SINCE!

Almost every Chevrolet engine that we build in our shop is equipped with the standard oiling system, modified to our clearances. Only the very hottest "all-out" competition engines are built with pressure oiling.

It is easy to see that the use of the stock Chevy oiling system will save you piles of money that can be used for a manifold, head, cam, or some other item of equipment that will do a lot more for your engine performance-wise. Certainly a pressure system of oiling will not do anything to make your engine faster in any way.

That too, is the primary reason why we used the splash oiling system on our dynamometer test engine, as we wished to prove the efficiency of the system beyond all doubt . . . Especially to those hard-headed die-hards who just can't look at the long record of successful use of this system in millions of stock Chevrolets and see how it can and does work in a high speed engine.

In several years of using the stock Chevrolet oiling system, the author has lost one rod bearing in one engine, and that was traced directly to the use of a widely advertised oil additive. In fact, when we suggested to a New York customer that he not use this particular product in his Chevrolet, and this word got to the New York distributor, the California distributor was contacted and told to have the author stop hurting the sale of this product. The California distributor of this very popular oil additive called on us and in very profane language

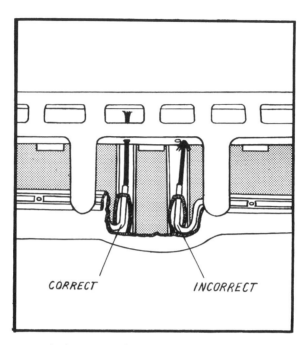

An oil pan target is shown in use here. Note that the oil spout which is set incorrectly does not send its stream through the proper hole in the target. The use of this target is recommended each time a pan is removed from a Chevrolet engine.

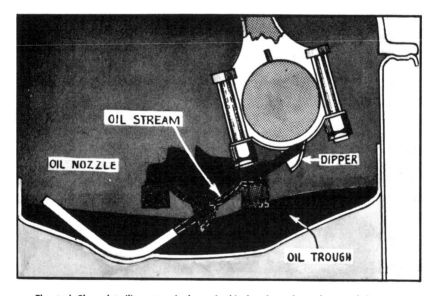

The stock Chevrolet oiling set up is shown in this drawing. As engine speeds increase, the effective oil pressure increases due to the streams of oil entering the dipper cups. This set up has proven very effective for even hot racing engines when set up with the proper clearances.

chided us for not using his product, as several race drivers and owners were supposed to be doing with great success. We indicated to the man that his product had caused harm in our engines. His offer was as follows: We were to build another engine, and use his oil additive. When the engine blew up (he did not say "if") we were to furnish samples of the crankcase oil and bearing material so that his firm might have them analyzed and tell us what had caused the failure. Thanks a lot—but NO THANKS. A bearing and oil analysis would be peanuts compared to the cost of a racing engine!

If a snake bit you once, you'd certainly not entice the reptile to do it again!

O.K., the data on clearances and set up contained in this manual are right Details have been shown so that you need not own a set of micrometers in order to get your clearances correct. Follow this information closely and you'll not have any trouble with rod or main bearings!

A final blast! Countless fellows have asked our advice on setting up rod and piston clearances, and we have always given them the proper information, as we have no "speed secrets." Yet, time after time, after one of these budding speed mechanics has built an engine—he comes to the shop and complains that his engine is tight, heats up quickly, perhaps freezes up, or has rod clatter after only 25 miles. When we ask if he used the clearances which we suggested, the stock answers are: (I) I don't know what clearance I used because I didn't have any micrometer and couldn't read it if I did! (2) Well, a friend of mine who is a mechanic told me that, etc., etc.

What a colossal joke on them! What a lot of fun they have paying for new rod bearings, regrinding the crank and so on! Would you take a sick dog to a butcher to find out what was wrong with him? Well then, why let some misled and uninformed Herkimer tell you how to foul up your engine? Will he be willing to pay for it if it blows?

It's a shame to have to use this space for a blast just to get across the idea that what has been proven right, IS RIGHT! Perhaps it will save a few engines and the name of the Chevrolet for racing. A few "zilches" working on their engines and pulling monstrous "foul-ups" which turn their engines into ill running "no-goes," can cause a whole section to go sour on the idea of hopping up Chevrolets or any other car.

Careful and precise workmanship costs less in the long run! Herkimers have no business working on their engines. This type of speed work is for the serious man who is willing to do careful mechanical work and use the correct tools and information for the job.

No Mikes?

No longer need you worry about getting rod and main bearing clearances correct, even if you don't own a micrometer of any sort. With this new product of PERFECT CIRCLE, it is possible to accurately check your clearances. The price is so low that no builder can afford to set up an engine by "hit or miss."

With PLASTIGAGE, it is only necessary to place a small strip of the plastic material on the clean bearing, tighten the bearing into position with recommended torque. Remove the bearing and compare the width of the plastic material with the scale found on the paper envelope in which the PLASTIGAGE is sold.

One packet of PLASTIGAGE is sufficient for the average engine. The plastic material is oil soluble and has been proven to be two-thirds faster than any other method of checking bearing clearances.

The principle on which PLASTIGAGE works is that the plastic material is flattened by the pressure of tightening the bearing cap. The less clearance there is, the greater the flattening, and the wider it will be. When the bearing cap is removed, the width is measured by direct comparison with the graduated scale on the PLASTIGAGE envelope. The numbers on the graduated scale indicate bearing clearances in thousandths of an inch.

1950-105 H. P. Hi-Torques

Several readers have expressed interest in installing these engines in their present Chevrolets, due to the improved design of the cylinder head and the large displacement of the hi-torque models.

The 1950 Power Glide and Hi-Torque models can be bolted right to 1941 and later bell housings. They are the same length and width overall, so that no installation difficulty should be experienced by any one making the change-over.

The 1950 hi-torque models have the solid tappets, and these should be used. The rocker arm assembly of the 1950 hi-torques and Power-Glides is different from the 1950 92 h.p. and 1941-49 standard and hi-torque models. A quick look at the head will also convince the most casual observer that the new cylinder head will not fit on the older models.

Therefore, it is not only necessary to buy a short block assembly (or the component parts thereof) to install this new engine, but also the rocker arm assembly and head, as well as the new short type push rod cover.

In the event that you should wish to install the 105 h.p. hi-torque in a pre-1941 Chevrolet (1937 or later) we would recommend that you use a late transmission and bell housing and install a column shift at the same time.

We often get letters asking if the Power Glide and 105 h.p. Hi-Torque heads will fit the earlier blocks. The answer is definitely "No," as this block and head assembly is an entirely new one. The larger intake valves were installed in this head in order to obtain further improvements in low speed torque so that the engine might better make up for the losses of the fluid drive and add low speed pulling power for truckers.

Powerglide Chevys

The Power Glide Models which were introduced on the 1950 Chevrolets have been ridiculed to a great extent due to the excessive gasoline consumption and lack of braking power.

Many of our customers have added speed equipment to these models with gratifying results. However, it is not possible to go as far with improving the performance on these models as can be done with those using the standard transmission.

This is due to the fact that the Power Glide is unable to transmit the horsepower to the rear end without loss of power and a large amount of slippage between the engine and the drive line.

In fact, there have been quite a few names coined for these models. Among these are: Power-Slide, Power-Slipper, and Power-Clipper.

One of our acquaintances has installed a very hot Chevrolet Power Glide engine in his Chevrolet with the fluid transmission and reports that the acceleration is not much better than that which could have been obtained by using the dual manifold, high compression head, and ¾ cam set up.

The use of a ¾ camshaft is permissable if you will remove the hydraulic tappets and install the old solid type. Hydraulic tappets can be used with hydraulic ¾ cam grinds. A more radical cam than this would undoubtedly create the condition known as creep, where the car tends to crawl or creep forward when the engine is idling, due to the fact that the more radical cams must be idled at higher RPM's in order to keep the engine running smoothly.

The use of an 8.5:1 high compression head is recommended on these engines as this will usually improve the fuel consumption and at the same time improves the all around performance of the car, including gains in top speed and acceleration.

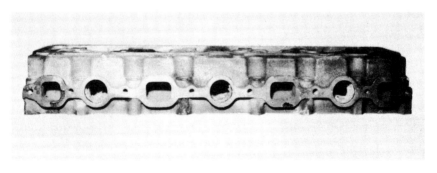

Side view, Power Glide cylinder head.

A NICSON dual manifold, installed with two ZENITH carburetors will also aid the acceleration, and will seldom cause any increase in gasoline consumption. This manifold can also be used with two ROCHESTERS but in this case it is wise to use the ROCHESTER carburetors supplied on the 92 h.p. Chevrolet engines as they have a smaller venturi size and are better adapted to a dual installation.

Definite improvement in the Power-Glide ignition system is possible through the use of the MALLORY two point conversion kit in the stock distributor, along with a MALLORY BEST coil. For a moisture proof, high quality distributor, the installation of the complete MALLORY two point Chevrolet distributor is recommended.

The installation of a set of hi-dome pistons in 1950-52 Power-Glides, or milling the head of the 1953's .080" (and sinking the intake valve seats a like amount—along with two carburetors and a suitable ignition will usually put the P. G. equipped Chevy in the 100 mph class.

Chevrolet ignition converted with Mallory two point kit. These are ideal for the average fast Chevrolet or GMC.

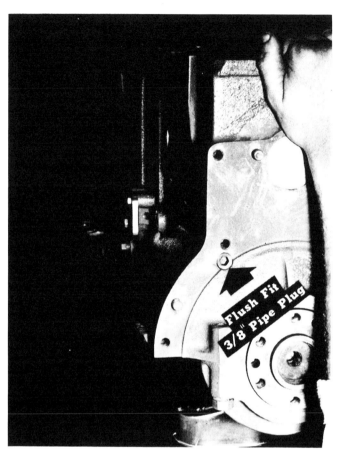

The front and rear of the Chevrolet main oil galley must be tapped for 3/8" pipe plugs. Allen plugs are preferred. Rear plug must be flush with block or bell housing will not seat properly.

Some of our readers are always interested in the details of installing a pressure oiling system in the Chevrolet engine. For long distance racing or track jobs where the oil sloshes around a lot, this type of oiling is ideal. For all out racing, this system is preferred by experts.

The installation of the pressure oiling system is easily accomplished and consists of few special parts. Foremost of the items needed for the installation is a drilled crankshaft. Your Chevrolet crankshaft must be drilled in a precision jig so that the oil holes will miss the forging holes in the crankshaft. A set of insert rods must be purchased, plugged for pressure oiling. Modifications of the block are simple and are clearly explained by the accompanying illustrations.

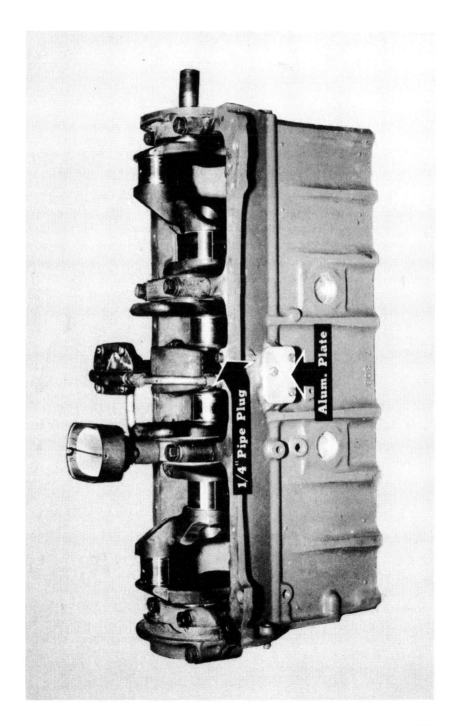

Chevy Pressure Oiling

To equip the Chevrolet block and crank assembly with pressure oiling it is necessary to:

1. Have your crankshaft grooved and drilled to provide passageways for the oil to travel to each of the connecting rod throws;

2. Before line-reaming the main bearings, tap each end of the main oil galley with a ⅜" pipe thread tap. The rear hole must be given particular attention so that the ⅜" pipe plug (preferably ALLEN type) which will be used to close the hole, will go into the block so that it is flush with, or slightly under, the edge of the block. This is important, as the bell housing must seat squarely against the block! (Consult the accompanying photograph to see which of the two holes in the lower flange of the block you will tap for quarter-inch pipe plugs.)

3. Blow out all passageways with compressed air;

4. Have the crankshaft ground, if necessary;

5. Install and line-ream main bearings with NO SHIMS. Clearance should be .0025".

6. Remove metal sleeve from block. This is the sleeve which carries oil to the spouts in the oil pan. This hole should be tapped for ¼" pipe and plugged (if a full-flow filter is not going to be used).

7. Insert-type rods should be used. If Chevrolet rods are used, then it will be necessary to tap the small holes in the upper half of the rod with 3/16-24 NS tap and install 3/16" x 3/16" Allen head set screws. Do not run the thread all the way through the hole or it will be impossible to get these screws in sufficiently tight.

8. Tap the holes in the bottom half of the rod with a ⅛" pipe tap. The hole should be tapped so that a ⅛" flat-head, screw-slot brass plug can be screwed into the rod far enough so as not to interfere with the insert. This hole should be tapped through from the inside of the rod. Prick-punch the brass pipe plug so that it will not screw out of position;

9. Grind off the protruding portion of the brass pipe plug which will be sticking out on the outside of the rod. (Rod bearing clearance should be .003" to .0035".)

10. A stock Chevrolet oil pump can be used and will usually provide 65-75 pounds per square inch of oil pressure if it is in good condition. Some mechanics prefer to install a G.M.C. oil pump, but this is not necessary, as the stock Chevrolet pump will provide sufficient pressure for the very hottest engines.

11. The rod dippers can be discarded and the spouts and troughs removed from the oil pan. Be careful when cutting these loose from the pan. It is easy to get cut on the jagged edges! Makes nasty cuts, too!

12. A pressure system must be primed before the engine is started for the first time. This is accomplished by inserting a length of drill rod with a flattened end, modified to fit into the driving slot of the oil pump. Chuck this rod in your hand drill. This must be done with the distributor removed from the block assembly. Of course, there must be oil in the pan. Turn the pump with the drill until oil pressure shows on the gauge, and oil appears around the rocker arm assembly.

13. The stock Chevrolet oil pressure gauge will not handle the pressure of your pressure oiling system. A 100-pound oil pressure gauge should be installed. We recommend a Stewart-Warner #D360A oil pressure gauge.

Stewart Warner D360A one hundred pound oil pressure gauge is used with Chevrolets converted for pressure oiling.

Gear Ratio

For drag racing over courses not exceeding ¼ to ½ mile in length, especially with a heavy bodied coupe or sedan, we suggest the 4.11:1 or 4.22:1 gear ratio.

If the power of your engine is such that you get excessive wheel spin in low gear, you may find that starting in second gear and using only second and high will give you better all around acceleration and a quicker time. This is common, particularly in Southern California, at many of the drag races. Often the cars which are going the fastest at the end of the course are using a low rear-end ratio along with using second and high gear in their transmissions. When high-speed cruising is the main purpose for which the engine has been built, we always suggest using the 3.73:1 ratio; or in extreme cases where the country in which the car is to be used is very flat, or no drag racing is anticipated, the 3.55:1 ratio can be used to advantage. Low speed acceleration will usually suffer when this ratio is used, unless you have an engine producing a great amount of horse-power. We used this ratio for obtaining top speed at the Dry Lakes.

If you wish to use your car for racing from standing, or barely rolling starts, the use of low gear instead of starting in second is to be advised. In this case, if your engine has such terrific power that it causes excessive wheel spin, it is advisable to change to a higher gear ratio or larger tires so that you will get better traction for take-off. We usually find that the best acceleration is to be obained in second gear when 3.73:1 or 3.55:1 raios are used.

John Lightfoot's Chevrolet powered roadster has turned 134 mph in Southern California Dry Lakes time trials. Engine uses three side draft Harley-Davidson motorcycle carburetors.

We are often approached for recommendations concerning the proper rear-end gear ratio to use in a particular Chevrolet. It is possible to install a gear ratio which is different from the one now in the differential of your car. Several gear ratios have been available for 1937, and later, Chevrolets, including 4.11:1, 4.22:1, 3.73:1 and 3.55:1.

The latter ratio made its appearance in the 1950 Power Glide models. It is possible to interchange these gears by making the changes which are shown in the table below. Chevrolet part numbers have been given for your convenience and guidance.

GEAR RATIO CHART

1937-39 Cars	602443 Stamped 593002-01	38-9	4.22:1
	602440 Stamped 595022-21	41-11	3.73:1
1940-51 Cars	604397 Stamped 595022 3652285	41-11	3.73:1
	604398 Stamped 3652178-79	37-9	4.11:1
POWER GLIDE	3694806 Stamped 3691464-65	39-11	3.55:1

As the intermediate ratios (low and second) of the Chevrolet transmission are quite low and have a large jump in ratio between them, the Chevrolet with a hopped up engine tends to "run out of gears" very quickly. In order to counteract this situation, several Southern California Chevrolet and GMC enthusiasts have equipped their competition cars with Ford gear boxes and Lincoln-Zephyr gears. A comparison of these ratios appears here:

	CHEVROLET	29 TOOTH FORD	28 TOOTH FORD	26 TOOTH ZEPHYR	25 TOOTH ZEPHYR
Low	2.94	3.11	1.607	2.327	2.12
2nd	1.68	1.77	2.82	1.576	1.435

With a fully loaded red hot engine, in our 1937 Chevy coupe, we were able to turn over 100 mph in second gear using the Zephyr 26 tooth cluster and 3.55:1 rear end gears. In order to use the Ford type gear box on the rear of the Chevrolet engine it is necessary to purchase an adapter bell housing which will fit the rear of the Chevrolet block as well as the Ford transmission. Also, the Chevrolet or GMC flywheel must be drilled to accept the   Ford large diameter pressure plate

At the present time, the only difficulty involved in making the change over to the Ford transmission lies in the universal-joint. On our coupe we used a take-apart Chevrolet rear half, with a take-apart Ford front half. We managed to blast three of these in six meets, which is a good average! The only really satisfactory way to make the conversion using the Chevrolet rear end is to get a needle bearing type Lincoln Zephyr U-Joint and get it re-splined to fit the Chevrolet drive line.

At least this conversion bell housing allows you to install Chevrolet and GMC engines in Ford type chassis, such as roadsters and early coupes converted for competition use. When using the Ford transmission with a Ford rear end, a Zephyr U-Joint should definitely be used as it is very strong and trouble-free.

1937-39 Chevrolet transmissions should be completely equipped with 1940 or later gears, including the needle bearing cluster gear. Some gear manufacturers here in Southern California make special short main shafts so that this is possible. This special main shaft is necessary to fit the synchro-mesh arrangement.

The stock Chevrolet (models 1940-51) transmissions are capable of handling many more horsepower than the stock engine produces. In over three years of hard use on the road, in drag racing and in city driving, the transmission of the author's 1941 Chevrolet coupe has experienced only one horrible blow up, and that was while shifting at full throttle from 45 MPH in low to second and listening for the fine sound of second gear rubber. In this particular instance all we got was a horrible crash; and climbed out to observe gears hanging out of the case at various angles while the oil dripped merrily onto the pavement!

1940-51 model transmissions are basically stronger than the 1937-39 models due to their stronger mainshaft.

If you plan to pour lots of horse-pressure into your transmission, 1940 or later model, it is advisable to change the bushing type cluster gear to a needle bearing cluster gear, as these will withstand the terrific strains imposed on the transmission when turning 60 to 85 mph in second gear. We have actually seen the bushing type gears turn blue from the heat which is generated. In order to change over to the needle bearing cluster, use the following parts:

1—Countershaft No. 591213, 7/8" x 7—7/32" with no flat spots.
1—Countershaft gear No. 591191 (cluster gear)
50—Rollers No. 435847 and 2—washers No. 591212

In order to install 1940 or later ratios in 37-39 rear ends, it is necessary to change the spline on the end of the drive line. This can be accomplished by cutting off the splined collar from the rear of a later drive line and welding this spline onto the rear of the early drive shaft. This should be done in a lathe so that all parts will be aligned properly.

1950-51 Power Glide Models are equipped with 3.55:1 ratio gears. In order to install this gear set in 1940-51 rear ends, it is necessary to change the case to which the ring gear is bolted. Part number for this differential case is 3691463, and it is stamped 593009. Under no circumstances should this gear ratio be installed without changing the case!

These four gear ratios allow a wide choice for varying uses, especially when various tire sizes are used. No other ratios are available for the Chevrolet passenger car rear ends at the time this book goes to press.

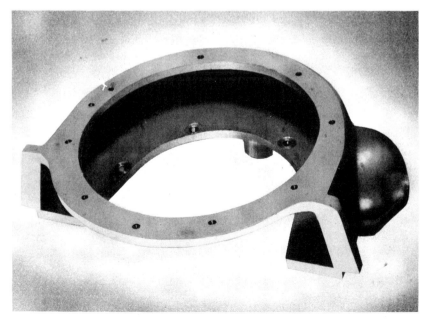

Cyclone Bell Housing for using Ford transmission with Chevrolet and GMC engines.

Flywheel Chopping

Reducing the weight of the stock flywheel is always a good idea with a hopped-up engine, as it serves to reduce the loading on the rear main bearing to a great extent.

Not only that, but it reduces the amount of weight which must be accelerated and decelerated when raising or lowering the engine RPM. However, this factor will cause the engine to idle a bit more roughly as it will not have the centrifugal force of the flywheel to carry the engine smoothly over rough spots.

In our opinion, the light flywheel also reduces the chances for blowing up a transmission or tearing out axles, as it does not give the tremendous surge (of the stock flywheel) when shifting gears. The stock flywheel will give as much as 7 to 10 mph "jump" when shifting from low to second gear. This jump is obtained at the expense of placing a terrific jolt of power on all parts of the driving train, especially the transmission and clutch Consequently, we now recommend chopping your flywheel for any job hotter than a semi road outfit.

After the flywheel has been chopped, it should be balanced with the pressure plate with which it is to be used. This insures that no vibration will occur from an out of balance clutch and flywheel assembly.

No more metal should be removed than is shown in the template we have reproduced here. Further chopping could seriously weaken the flywheel so that it would fly apart at high speed and cause really serious harm to you and your machine.

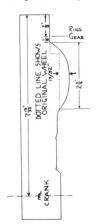 

Shown here are a chopped Chevrolet flywheel and a template which are used when the flywheel is turned in a lathe. No more metal should be removed than is indicated by the template. This insures adequate strength for the flywheel.

How Far Shall I Go?

SEMI-ROAD JOB (Gives 25-35 horsepower over stock)

1. ¾ cam
2. Dual manifold, or stock
3. Milled head
4. Aluminum pistons, .030" over torque for 37-40 models, .060" over torque for 41-51 models, standard or hi-torque. Fit with .0025" to .003" clearance.
5. Lower end will be O. K. if in good condition. Check clearances with Plasti-gage to make sure.
6. MALLORY 2-point ignition kit.

HOT ROAD JOB (Gives 50-75 horsepower over stock)

1. Full or super grind cam
2. Dual or triple carburetion
3. Aluminum split-skirt pistons fit .0035" to .005" clearance, bore as in semi-jobs. Hi-torque can be bored to 3-21/32" (.090 oversize). If solid-skirt pistons installed, bores larger than these not recommended as they weaken the cylinder walls
4. .003" to .0035" clearance on insert type rods, mains line bored with no shims to .0025" clearance
5. Milled, filled and ported head, 8.5:1 or higher compression. Ratio will depend on type of fuel you plan to use.
6. Aluminum timing gears
7. Chopped flywheel and heavy duty clutch
8. Chevrolet outer, special inner springs on intake Chevrolet outer, Buick inner springs on exhaust
9. Competition retainers on intake
10. Tubular competition pushrods
11. Oversize intake valves for acceleration or drag racing
12. Splash oil system O. K. (if properly set up)
13. CALIFORNIA BILL "Special" Distributor
14. Dual mufflers and headers should be added for best performance

ALL-OUT ENGINE (100-125 or more, horsepower than stock)

1. Hi-torque block bored .090" or Standard bored to 3⅝". Solid-skirt pistons with .990" pins
2. Pressure lubrication to GMC type rods
3. Triple side drafts, down draft carburetion, or fuel injection
4. Exhaust headers
5. SCINTILLA Magneto,
6. Super cam grind
7. Rest of equipment as for Hot Road Job
8. Should be equipped to use alcohol fuel, if permitted
9. Degreed crankshaft pulley.

Oil

Oil is important in your engine. The selection of a good grade of oil, as well as the proper weight, will do much toward the continued long life of your high-speed Chevrolet engine. As we have used both VALVOLINE and HAVOLINE motor oils with great success in the engines which have been built in our shop, we can highly recommend these brands. Most of our racing engines for road use are broken in using S.A.E. No. 20 oil; and the continued use of that weight of oil is advised until signs of using oil are evident, at which time a change to a heavier grade of oil is recommended. In some of the colder areas, an engine which is new should certainly be broken in with 10-weight oil, especially during the winter.

Oil additives are not in as good favor with race owners, drivers, and mechanics, as current advertising might indicate. We have used several makes of oil additives and at this time can place our stamp of approval on RISLONE, MIRACLE POWER GRAPHITE OIL, and a product made locally in Southern California called SOLV-X HI-TORQUE ADD OIL. We have had good results with these particular brands.

How To Rework Your GMC Engine

No longer do Chevrolet owners have an excuse when some owner of a "3/8 x 3/8" Mercury "shuts them off" at top speed or in a drag race. Heretofore, the Chevrolet owner has been restricted to approximately 250 cubic inches as a top displacement limit which he can obtain from a hi torque block, and less than that with the standard passenger car block. Now you can own a Chevy with an engine of 300 cubic inches displacement if you so desire!

Fortunately, the G.M.C. truck engine has come to be well regarded as ideal for racing purposes. Why the builders of high speed reworked U. S. engines overlooked these engines as long as they did is hard to understand, as they are perhaps the easiest to adapt for high speed, and certainly one of the most suitable engines on the market, as well as inexpensive when cost per horsepower is considered.

The G.M.C. engines are almost everything (stock) that the Chevrolet can be built up to with a lot of work. They have a very heavy duty oil pump, with a fine pressure oiling system. The design of the overhead valve mechanism is far advanced. Recently the G.M.C. factory has produced a new cylinder head to fit all of the Group I engines. It has such terrific port area and advanced design that it must be seen to be appreciated.

By mentioning Group I engines, the author has slightly "jumped the gun," for many readers may not be familiar with the classifications of the G.M.C. truck engines. A brief breakdown of the engine sizes and groups may be helpful, especially to the many readers who are going to become "Jimmy" minded after reading this manual.

The Group II & III engines are quite heavy, not only that, but their physical size, as well as huge displacement, renders them impractical for racing purposes. Race car limitations as to cubic inch displacement will place the II & III engines out of the discussion which is to follow.

We shall be primarily interested in the Group I engines as they are most suitable for racing, and are also easy to adapt to the Chevrolet chassis.

The 228, 236, and 248 engines have a 3-13/16" stroke, while the 256 and 270 engines have a 4" stroke. Each crankshaft is interchangeable with the other, and the rod and main bearings are the same size. While the blocks are basically the same in outward appearance, the cylinder wall thickness varies in the different models.

The 270 block, which comes stock with the 4" stroke crank, can be bored out to 4" which gives more than 300 cubic inches of displacement (301.6 to be exact!). If the bore is held to 3-31/32", the displacement will be 298.2 cubic inches, which will place the engine under the Dry Lakes competition rule limits as to displacement as they stand at present.

The top or face of the G.M.C. 270 block, and the 256 models is 3/32" higher than the other Group I engines, so that the same connecting rods may be used throughout. Alert Chevrolet enthusiasts will note that this is the same engineering stunt utilized on the hi-torque Chevrolet blocks so that the same connecting rods may be used on both passenger and hi-torque blocks.

Group I cylinder heads will interchange and the new 1950 — 270 GMC heads will fit all Group I engines. These heads have 1-3/4" intake ports, being 3/8" larger inside diameter than earlier models and thus effecting an increase in port area of 62% over the older heads.

LOWER END

The rod and main throws of the GMC crankshaft are the same size as those used on the Chevrolet, but the wristpins are of a much larger diameter (.990") which renders excellent stiffness for the piston. Many racing engineers feel that the large diameter wrist pin is a large factor in keeping pistons from breaking up in a high speed engine.

The inserts used in the main bearings of the GMC are precision type, and if purchased in the proper size should not require any line boring to insure accuracy. The author recommends that you grind your GMC crank to a standard undersize plus an additional thousandth under size so that there will be ample clearance. This should be done on both rod and main throws. For instance, if the crank will "clean" at .001" undersize, then have it ground that way and use standard bearings. If it will clean up at .010" under, then have the shaft ground .011" under straight through.

If you intend to line bore your GMC block, then it will be wise to purchase .020" under main bearings (sizeable to standard). If your crank has been ground .010" under, then buy .030" undersize bearings, sizeable to standard.

Forty-five to fifty pounds of oil pressure are provided by the stock GMC oil pump. This is ample for a hot road job. If you desire more oil pressure, then use the next heavier wire gauge spring in the oil pump relief valve, which will provide 80 pounds oil pressure (approximately).

Use the oil pump with the stationary oil screen. Be careful not to put flat washers behind the relief valve to increase pressure. This will interfere with the full travel of the plunger, and could block the by-pass sufficiently to create a pressure of 800 to 1000 pounds oil pressure with a cold engine!

Connecting rods of the Group I GMC engines work out fine, and no modifications are necessary. Check the wristpin bushings to make sure that they are not sloppy. They should be fit loose enough to allow the wristpin to slowly travel through the bushing when the rod is laid on its side, resting against the extended end of the wristpin.

For a road job it will not be necessary to groove the main bearing journals of the crankshaft. For racing, the grooving of these journals will greatly aid the oiling available for the rods at high speeds. These grooves can be made in the mains for you by your local crankshaft grinder. Cut the grooves so that they will coincide with the holes in the crank. The groove in each journal should be approximately 3/32" to 1/8" wide, and about the same depth. This grooving should be done before the crank is ground, as it occasionally causes a slight warping of the crankshaft.

DISPLACEMENT CHART

BORE	STROKE	CUBIC INCHES	REMARKS
3- 9/16	3-13/16	228.0	GMC "228"
3-23/32	3-13/16	248.5	GMC "248"
3-25/32	4	269.5	GMC "270"
3-29/32	3-13/16	274.1	Horning "Indianapolis"
3-15/16	3-13/16	278.5	Optional
3-15/16	4	292.5	Recommended for 270 Block
4	4	301.6	Has Been Done

THANK YOU, ROAD & TRACK MAGAZINE

It will probably be necessary for you to place baffles in the bottom of the pan, especially from the rear half to the rear of the engine. One baffle (liberally pierced with holes) should be placed horizontally. These engines have such terrific acceleration when set up properly, that the oil tends to pile up at the rear of the pan when accelerating and thus starves the oil pump, dropping the oil pressure below the danger point.

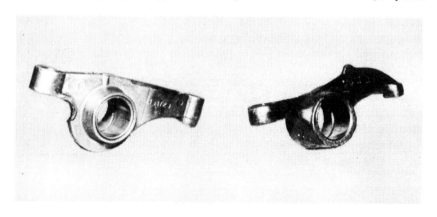

Cast iron and new fabricated type GMC rocker arms. The fabricated type are stronger and lighter than the cast models and should be used for racing engines.

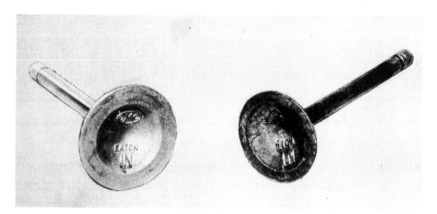

Comparison photograph of 1950 - 270 GMC intake valve with earlier Group I GMC intake valve. Special attention should be paid to getting seat to outer edge of valve for maximum valve area.

GMC ENGINE GROUPS

Group I	228 - 236 - 248 - 256 - 270
Group II	278 - 308
Group III	361 - 426 - 477 - 503

CYLINDER HEAD

We highly recommend the use of the 1950—270 GMC cylinder head, as its port area is approximately 62% greater than earlier models. If you are building a mild road job and already have the older model head, it can be ported out to the flanges in the side of the head, and careful enlargement of the intake ports can be effected by use of a grinder.

It is not recommended that you improve or increase the compression ratio by milling the head. Instead, high dome, high compression pistons should be used to get gains in the compression ratio. When using either the 1950 or earlier heads, the seats of the intake valves should be narrowed so that a larger area will be available for the fuel to enter the cylinder when the valve is raised from its seat.

Valve spring combinations will be almost identical to those used for the Chevrolets. Chevrolet outer and special inner will be strong enough for nearly all competition use. The use of the heavy combination just mentioned tends to increase valve noise. It is recommended that you use a less stiff combination if the cam you buy will allow it without causing valve float.

Either competition spring retainer seats or the 1950 Chevrolet spring retainers can be used without fear of pulling the valves through the retainers at high speed. Under no circumstances should the pressed metal retainers (stock type) be used, as the loss of a valve at high speed can really tear up your engine before you realize what is happening!

The springs should measure 1-27/32" on the seat, that is, measuring from the base of the spring to the top of the spring, with the valve on the seat and the keepers and spring retainer in place. If your springs show less height than this, the seat must be sunk deeper into the head. If they show more height than the recommended figure, shim them up by placing 1/32" or 1/16" washers under the springs.

1950 sheet metal, fabricated rocker arms should be used for an "all-out" GMC competition engine. The cast or forged models will be sufficient for a road engine. BARKER Hi-Lift Rocker Arms can be used successfully on the GMC engines, as the GMC and Chevrolet engines use identical intake rocker arms.

The cylinder head must be installed using a torque wrench. 85 to 95 foot pounds of torque should be used.

PISTONS

Two pistons are in wide use in converted GMC engines. Both are made by Frank McGurk. One is the Venolia solid-skirt type for all 270 GMC's. This piston provides 8.2:1 compression, and is available in 3-29/32, 3-15/16, 3-31/32, and 4" bore sizes. The piston for the new 302 GMC is a slipper skirt flat top job now available in 4-1/8 and 4-3/16 inch bore sizes. A 4-inch piston for 302 is made by California Bill.

3-15/16 inch is the recommended bore size for 270 GMC's, as this leaves room for at least one, and perhaps two rebore jobs. The 4⅛-inch bore is recommended as maximum for the 302 since this is the largest bore that the stock gasket will clear.

PUSHRODS AND VALVE TAPPETS

Tubular pushrods must be used in the GMC racing engine, as they are lighter than the stock pushrods, and therefore have less tendency to cause valve float. They will not flex or bend under the additional valve spring tension which is necessary.

For valve tappets (cam followers), purchase old ones from the junk yard. The 1937 Chevrolet cup type will fit the GMC. The old ones that have been run in engines will be glazed from use, and all of the pores of the metal will contain grease and carbon. The faces of old tappets are far harder than new ones, and to use new ones is false economy as they will not last half as long as an old set.

This GMC engine powered California Bill's 1941 Chevrolet coupe to a speed of 126 mph. Equipped with McGURK pistons, NICSON manifold, ZENITH carburetors, and SCINTILLA magneto. Special late model 270 H Cylinder Head was equipped with 1-11/16" exhaust valves. Late model rocker arms were actuated by HOWARD F-7 camshaft, HOWARD pushrods. Alcohol fuel was used.

TIMING GEARS & CAM TIMING

Cam timing on a GMC engine is of utmost importance, even for road jobs running a mild ¾ grind. To check the cam timing requires the use of a degree wheel, and preferably a dial indicator. You will also need the timing or checking clearance of the cam, or your check will be meaningless. If this information is not included on the tag with your cam, you will have to get it from the cam grinder, as this varies with each particular make and grind of camshaft.

Although stock GMC crank gears vary widely in their timing, the camshaft gears are quite accurate in most instances. For this reason you may find it necessary to experiment with several crank gears before you find one that can be jockeyed to give the right timing. Above all, don't try to interchange Chevrolet and GMC crank gears. Even though they look alike, they are different by ½ to a full tooth and their keyways are cut in different relationship to the gear teeth.

WATER PUMP & CRANK PULLEY

Use a 1939 Chevrolet and water pump with your Chevrolet generator and crankshaft pulley. It will be necessary to cut ¼ inch from the shank of the Chevrolet pulley at the point where it butts against the crank gear, otherwise the belt will not align properly. This cutting can be done in a lathe. Use a GATES 3440 Fan Belt.

STARTER AND GENERATOR

We suggest that you enlarge the mounting flange holes on a Cadillac starter and use this on your GMC, especially if you are running very much compression. These starters have quite a bit more cranking torque than the Chevrolet starters.

The stock Chevrolet generator fits right onto the GMC mounts.

CLUTCH

Competition and dynamometer tests, under varying conditions, have shown that it is necessary to use the large ROCKFORD clutch with 12 125-pound springs, giving a 1500 pound clutch. This is the large, truck model, and it should be used on a '38 or '39 Chevrolet truck flywheel. The stock diaphragm clutch is useless with a GMC engine as it will not handle the tremendous horsepower which these engines provide at the flywheel.

Close up of flywheel bolts on GMC. Four additional bolts must be installed between the stock bolts, making a total of eight in all. 7/16" SAE thread Chevro'et Flywheel bolts are preferred for the extras.

INSTALLING GMC ENGINES IN CHEVROLETS

This is perhaps one of the easiest engine conversions that can be made. Certainly it is easier than building the common A-V-8, so popular throughout the United States.

1937 through 1948 Chevrolets can be equipped with the GMC engines quite easily. The necessary modifications follow. Make sure that the bell housing used on the engine is 1938 or later. The radiator must be moved forward by two inches. We usually move the radiator to the other side of its mounting frame to obtain the needed location. The front motor mounts must be also moved forward by approximately 1½" so that they will line up with the front motor plate of the GMC. It is usually necessary to cut away a small portion of the front cross member to clear the GMC oil pan. This will be readily apparent.

1949-51 models require a slightly different treatment, due to the fact that they are equipped with a different type of steering arrangement with a center pivot arm right under the oil pan. In fact, this pivot arm must be dropped out of the way before the stock oil pan can be removed on these models. Naturally, the large GMC oil pan will not fit over this pivot arm without causing interference, so it is necessary to cut the pan to clear the pivot arm, as shown in the photograph. The pan should be cut at the front, right back to the forward edge of the oil pump. Allow about ¼" to ½" clearance at the bottom of the rods when welding the bottom of the pan in its new position.

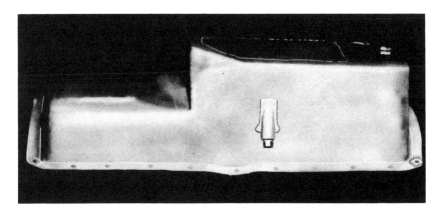

This GMC oil pan has been reworked so that it may be installed in 1949 through 1951 Chevrolets. Pan must be reworked as shown here in order to clear the center pivot arm of the steering assembly.

The rugged lower end of the GMC engine is pictured here. This lower end will withstand horsepower outputs in the neighborhood of 300 H.P. without strain.

With the 1949-51 models, the radiator should be placed on the other side of the mounting frame, and it may be necessary to lower the radiator somewhat to clear the hood line.

While this conversion takes a bit of time to do correctly, the horsepower produced by the GMC engines, coupled with the lightness of the Chevrolet chassis, produces truly amazing performance!

Holding No. 1 in the Eastern Oregon Racing Association for over three years, Mr. Rolla Vollstead's car is equipped with magnesium wheels made from his own pattern. GMC powered, the car has proven its worth on countless dirt and asphalt tracks. Photos shown here of its engine indicate the FISHER head with VACTURI carburetors converted for floatless operation. New equipment now in use includes a HILBORN injection, FISHER accessory drive and SCINTILLA magneto.

270 GMC engine in 1950 Chevrolet coupe chassis. Fram oil filter was installed on rocker arm oil line. Engine has Super camshaft, Nicson dual manifold with Zenith carburetors, Mallory ignition and Stewart Warner Tachometer. 292″ displacement, 8.5:1 compression.

Keith Loomis' GMC-Zephyr has 292″ displacement, 9:1 compression ratio. Manifold is home made, used Rochester carbs. These have now been replaced with Zeniths. Cam is Super grind, two point, dual coil ignition is used. Car is a Zephyr 1937 coupe.

GMC Set Ups

G.M.C. HOT-ROAD JOB

1. Full race cam
2. Early or late head. Early type head should be equipped with '50 intake valves
3. Dual or triple carburetion
4. MALLORY Distributor and Coil,
5. Chevrolet outer, Buick inner springs. (Make sure Buick inners don't bottom with cam at maximum lift)
6. Crank ground to standard undersize, plus .001" additional undersize (for clearance)
7. Racing pistons, Sealed-Power Rings
8. Degreed crankshaft pulley
9. Tubular pushrods
10. Aluminum Timing Gear
11. 1500 lb. Rockford heavy duty clutch.

ALL-OUT RACING ENGINE

1. Crank ground as for Hot Road Job and Groove Crank
2. 1950 Style G.M.C. head, ported. Give special attention to width of valve seats.
3. Competition retainers, inner valve springs
4. 10:1 or more compression
5. 3 side draft Zenith carbs, 3 Vacturis, fuel injection, or down draft 3-carburetor manifold.
6. Degreed pulley
7. Tubular pushrods
8. Chevrolet outer valve springs, special inner
9. Sheet metal rocker arms (late G.M.C.)
10. Sealed-Power Rings
11. 3-15/16" bore, 4" stroke to get 292 cubic inches, or maximum displacement recommended
12. Racing pistons with 10:1 compression ratio.
13. Equip for alcohol fuel use
14. Chopped flywheel, heavy duty, large ROCKFORD clutch.

Clutch here is ROCKFORD for Chevrolet passenger car. Models are available with beefed-up spring tension for high horsepower engines and severe racing type use. Larger models are available and recommended for use with GMC engines.

ASSEMBLY DIFFERENCES, G.M.C.

Assembly of the G.M.C. engine is the same as the Chevrolet with few minor exceptions:

1. Rocker arm oil supply comes from a fitting at the front of the main oil galley and is led into the front right side of the head. Rocker arm stands are each secured by two bolts.
2. Head bolts should be torqued down to 90 pounds as with the Chevrolets. Rod nuts need torque of 45-50 lbs. and the main bearing bolts should be torqued down to approximately 100 ft. lbs.
3. We do not usually line bore the G.M.C. blocks, as the precision main bearing inserts of the G.M.C. line up well with no shims required.
4. Crankshaft should be ground to a standard undersize plus an additional thousandth undersize. This applies to both rods and mains.
5. For rod and main bearings we prefer MORAINE bearings, but McQUAY NORRIS or FEDERAL MOGUL can be used successfully.
6. When assembling pistons to rods, wrist-pins should be fit with palm press fit. Install wrist-pin locks to secure wrist-pin in piston, or if desired, aluminum pin buttons can be used. Where pin buttons are utilized, no pin locks are used.
7. If installing pistons with partial domes, such as ½ domes, the pistons should be assembled so that the low sides of the pistons will face toward the spark plugs.

Dick Winfield of the Glendale Coupe and Roadster Club is shown here as he adjusts fuel mixture of his GMC engine. Mixture adjustment with the HILBORN FUEL INJECTION is made by varying the amount of fuel by-passed from fuel pump to fuel tank. Note special drive used instead of stock timing cover.

Winfield GMC Coupe

Throughout 1952, this 1932 Ford coupe dominated its class in Southern California drag racing, primarily due to its fantastic low speed torque and ability to "get out of the chute" long before other cars got under way. This particular trait enabled the winning car to turn in faster elapsed times, although its top speed at the end of the quarter mile was often bettered. Equipped with a FISHER head, this GMC 270 engine has been bored to 3-15/16" making its displacement a total of 292 cubic inches. Other equipment includes HOWARD F6 camshaft, SPALDING ignition. A 10.5:1 compression ratio is used with methanol fuel, blended with nitro-methane, and Champion LA-15 spark plugs. Rear end is a quick-change HALIBRAND.

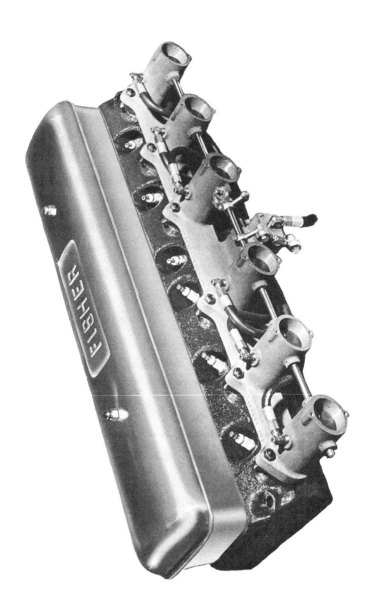

Weighing only thirty pounds, the FISHER 12 PORT GMC HEAD (Wayne F. Horning designed) boasts other aluminum advantages including freedom from hot spots and ability to allow more compression with gasoline, or more nitro-methane (up to 50%) with methanol fuels. Spark plug is located close to center of chamber, eliminating need for excessive spark lead. Port layout requires billet camshaft. Two-piece rocker cover eliminates oil drippage during tappet adjustment and aids general engine cleanliness.

Don Montgomery's Hudson-Buick

Many drag artists have been in for the surprise of their lives when Don Montgomery came to call in "my father's Hudson" as he puts it This sedan has "shut-off" innumerable roadsters which have turned 120 mph at the Dry Lakes, and the sedan itself has clocked an amazing 118 mph during the 1950 season of the Russetta Timing Association.

Don and the author are both members of the Glendale Coupe and Roadster Club, and as such participate actively in competition events in this area.

The car is equipped with a Buick Roadmaster engine, bored .125" oversize and equipped with solid-skirt, racing turbulator pistons. The head has been ported and all parts of the valve mechanism completely reworked to obtain maximum reliability with minimum weight.

A HOWARD camshaft has been installed and provides astounding acceleration. When "Monty" hits second gear his big sedan seems as if the front end will lift right off the ground. It gives a big leap forward and really starts to haul.

Carburetion is supplied by four STROMBERG 48's burning alcohol. Of course, Nitro-Methane is included as an active ingredient in the fuel. These carburetors are mounted on individual necks, which are balanced to some degree by coupling ½" neoprene tubing between the necks to secure equalization and keep the carburetors from popping back at low speeds.

Individual headers for each port have been made of flexible tubing. Ignition is provided by a converted PIERCE-ARROW dual 8 distributor. A photo of this distributor has been included for Buick fans.

This car was originally run on the road as transportation, but is now towed to and from events in which it competes so that it can be kept set up for racing. Changing back and forth gets to be a tedious process when the car is used for lots of competition.

This "pelican" (as we affectionately call it) has been clocked at 98 mph at the end of a rollingstart quarter mile at the Santa Ana Drag Races. This time has secured the trophy for "Heavy Sedans" on many occasions for Mr. Montgomery.

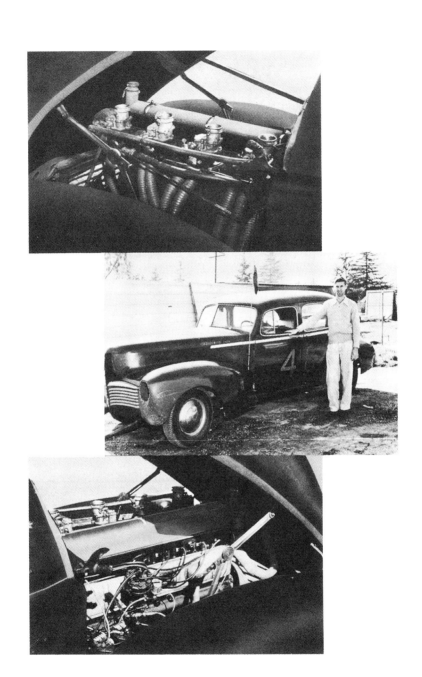

Top photograph shows the four carburetors used on Montgomery's Buick engine in the Hudson chassis. Middle photo shows car with owner standing beside it. Lower photo shows Pierce-Arrow dual coil, dual point ignition on right side of engine. Car is equipped with Howard cam.

How To Rebuild Your Buick For Speed

Let's face it! The average Buick engine does not show much promise in its stock condition. Its acceleration is certainly not extraordinary and its top speed, by actual dry lakes electronic speed timing, is approximately 90 MPH. If, however, you are the proud owner of a Buick and would like to be prouder yet, a few simple modifications will serve to greatly improve its performance.

Apparently not much was done with hopping-up Buicks until just after World War II. Recently, the author has noticed that sports car fans have used these engines to advantage in foreign sports car chassis. Not only that, but especially in Southern California, Buick owners, tired of being "blown-off," have added speed equipment and secured performance which is really hair-raising, as compared with that available from the stock Buick engine.

The Buick is a reliable engine and many motor rebuilders have indicated to the author that, if all engines were to have the same stamina and longevity as the Buick, there would be a lot less motor rebuilding necessary.

Many Buick owners report as much as one hundred thousand miles, with the boast that the head has never been off their engine. While we do not recommend this as an ideal practice, it certainly indicates the long lasting qualities of these fine engines.

Since it has often been said that there is no substitute for cubic inches and that added displacement is the easiest way to gain horsepower, it is quite obvious that, if at all possible, you should use the large Buick engine. These have 320 cubic inches as they come from the factory. These large engines have a bore of 3 - 7/16" with a stroke of 4 - 5/16". These engines are used in the Century, Roadmaster, and Limited Buick models. The smaller 248 cu. in. engine, with a bore of 3 - 3/32" and 4 - 1/8" stroke are found in the Special and Super models.

1950 saw a new Buick engine model offered to the public, which was designed by the factory as the "F-263." This is a bored-out 248 cu. in. engine with a 3 - 3/16" bore and the stroke remaining at 4 - 1/8". This would indicate that the cylinder wall thickness of the smaller Buick engines is quite adequate for large, oversized bores.

As with the Chevrolet and G. M. C. engines, we will attempt to break down the various modifications into classifications which will enable you to go about your souping-up operations in an easy and orderly manner.

LOWER END

The Buick has a positive pressure oiling system which, with the stock pump, provides 35-40 pounds pressure at 40 MPH. No modifications to your oil pump should be needed if it is approaching this pressure and is apparently in good condition. The author highly recommends that you install 1949 or later connecting rods, as these rods are designed for insert bearings. These later rods are interchangeable with the earlier babbitted models. The thin layer of bearing material backed with steel, as is found in precision inserts available today, will carry greater loads than plain babbitt. It is easy to see that your rod bearing life will be longer and give less trouble when the insert-type rods are used. Connecting rods should be set up with at least .003" clearance or as much as .0035" clearance. Less will ruin a set of rods in short order.

The Buick block is equipped with five main bearings of the insert type. It is strongly recommended that you purchase these bearings in an undersize and have them line-bored by your local auto machine shop to insure absolute accuracy of all the bearing surfaces. No shims should be used in the main bearing caps when you have the job line-bored. You may have a small fight on your hands when you try to get your local auto machine shop to do it this way, but we have found that it is always best to install main bearings with no shims, line-bored to the proper clearance. Replace the main bearings when they become excessively worn, instead of trying to remove shims, which will not give you an even bearing surface around the diameter of your main bearings. A clearance of .0025" is recommended for the main bearings. It is highly recommended that if your crankshaft is in need of regrinding, you have the rod throws ground to a standard undersize, plus an additional thousandth undersize. Then a standard undersize insert can be used and still give you adequate clearance. Clearance can be checked by using either a micrometer or PLASTIGAGE, as has been illustrated elsewhere in this manual.

Crankshaft end play, measured between the middle main bearing flange and the crankshaft, should be .004"—.008". This clearance can be measured with a feeler-gauge inserted between the thrust edge of the bearing and the crankshaft, with the crankshaft pushed as far as possible toward the front or rear of the block, depending on which side of the thrust bearing you are measuring.

PISTONS

Pistons are available in standard oversizes in both split-skirt and solid-skirt type. Solid-skirt racing pistons are recommended for an all-out racing engine. The split-skirt type will work quite nicely in even the hottest road job if the pistons are checked occasionally to make sure there are no breaks or cracks where the head of the piston joins the skirt. The wrist pin in a Buick piston is secured to the connecting rod by means of a clamping arrangement tightened by a bolt. This bolt should be tightened to thirty-foot pounds torque with a torque-wrench. When tightening these bolts, it is strongly recommended that a drift, or long punch, be placed securely in a vise and that the wrist pin be placed over this drift while tightening the wrist pin clamp bolt. This will insure that the rod is not twisted out of alignment by the twisting action on the bolt. If you have access to an alignment gauge, the piston and rod alignment should be checked. This is especially important due to the long stroke used in the Buick engine.

Apparently two types of pistons with the Turbulator dome tops are available. One of these has a recess in the Turbulator area and this type has been used on all models, 1941 and later. 1938-40 Turbulator pistons will give a slight increase in compression ratio, since the Turbulator on these pistons has a flat area, where the later models are recessed. The solid-skirt pistons mentioned previously are available with the Turbulator heads.

It is suggested that a split-skirt type piston be fit with a clearance of .003" - .0035". Solid-skirt types are supplied with the proper clearance when you indicate bore size with your order.

The small Buick engines may be bored .125" oversize without danger of going into water or weakening the walls excessively. We would not recommend boring over 3 - 7/32" on the small 248" or F-263" engines and, while some have bored out the larger 320" engines to as much as 3 - 9/16", we would not recommend boring over .100" oversize except where an all-out racing installation is contemplated. .125" oversize bore on the 248" engine will provide you with 268 cu. in. The same overbore on the 320" engine will give you 343 cu. in.

GRANT piston rings are recommended and have been used with great success in reworking Buick engines. Allow .003" ring-gap per inch of bore diameter. For further information on ring installation, it is suggested that you read the information on setting Chevrolet rings.

CAMSHAFT

Perhaps the greatest improvement in performance available to your Buick can be obtained by the installation of a reground racing-type camshaft. Three-quarter, Full and Super grinds are available. Wherever possible, it is best to secure the advice of someone who has run these engines so that you will be sure to obtain a camshaft which will perform to your satisfaction. Special camshafts are available for the hydraulic tappets, but it is strongly recommended that the reader not give any serious consideration to using the hydraulic tappets in an engine which is expected to produce terrific acceleration and high speed. This is due to the fact that the hydraulic tappets will not follow the cam action at high speeds. If your Buick engine is equipped with hydraulic tappets, it is advisable to replace them with solid Buick tappets. Many reground camshafts will require an increase in spring tension. We do not deem it advisable to shim-up the valve springs to secure more tension due to the fact that Buick inner valve springs will become coil bound (squeezed solidly together) when a large shim is installed under the spring and a hi-lift cam is used. We would suggest, instead, that you leave your spring set-up absolutely stock for the average 3/4" grind and, if a Full or Super cam is used, that Chevrolet outer valve springs be installed in place of the stock Buick outer springs and that the Buick inner spring be left in place. If you have a comparatively new engine, it is highly recommended that you secure an old set of valve tappets (cam followers) from an old Buick engine at your local junk yard. It would be advisable to check on the current price of these followers and not pay more than half the list price to your junk yard. It is always a good plan to check on the current list price of any item before trying to buy it from a "junkie." Otherwise, you may find that you have paid more than list price for junk.

It is imperative that old followers be used, as these are coated with a glaze of grease and carbon which will greatly add to their life in your high speed engine. If your present set of followers is apparently in good shape and not nicked or galled, then they may be used.

CARBURETION

Carburetion does not present any terrific problem with the Buick engine as both commercial and factory made manifolds are available and can be used satisfactorily. Two carburetors will prove adequate for the hottest road job, but where severe competition conditions are to be encountered, it is highly recommended that your Buick engine be equipped with four carburetors designed or converted to use alcohol fuel.

For further information on fuels, we have prepared a chapter in this manual for your convenience and reference.

1941 and 1942 Buick engines were equipped with what the factory called "compound carburetion." The factory set-up was designed so that only one complete carburetor was used and this was on the front flange. The carburetor mounted on the rear flange contained only an idle and main jet set-up, with no accelerating pumps. The manifold was designed so that the rear carburetor was opened only at high speeds by a special damper valve.

In order to convert these stock manifolds to satisfactory dual carburetor operation for use on your hot road engine, it is necessary to remove the damper valve from the rear flange and install a carburetor identical to that used on the front flange. The carburetors must be synchronized so that they will open and close together. It is highly recommended that a ball-joint type of linkage be used to insure a smooth, non-binding operation of the carburetors. Many Buick owners have used carburetors from the smaller Buick engines when installing dual carburation on the 320" engine, and STROMBERG 97 Ford carburetors when installing two carburetors on the smaller engines. However, if at all possible, it would be exceedingly desirable to install two ZENITH dual throat carburetors, as these, with their wide range of venturi sizes, as well as many sizes of power, main and idle jets, will serve to greatly lessen the difficulties of getting the proper mixture at all speeds.

IGNITION

Two types of ignition are available for the Buick owner desiring to improve the performance of his ignition system. The installation of a MALLORY distributor, with a MALLORY "Best" coil, will provide a great improvement in ignition performance. For owners desiring the very finest for their engines, it is suggested that they install a SCINTILLA VERTEX Magneto. Either of these ignition systems will be far superior to the stock Buick distributor and coil set-up.

CYLINDER HEAD

The ports in your Buick cylinder head can be enlarged to the size of the counter bore which was placed there by the factory for use with the manifold aligning rings. The valve seats should be modified so that the intake valves will seat with the outer edge of the seat approximately 1/32" - 1/64" from the outer edge of the valve. When this has been done (for further details, see the information on porting Chevrolet heads), the valve pocket can be ported to the inner diameter of the valve seat.

These modifications should be carried out on both the intake and exhaust valve pockets. The intake valve seat should be narrowed to approximately 1 16" - 3/32", while the exhaust valve seat width should be 3/32" to 1/8" wide.

EXHAUST

The exhaust manifolds which are furnished with the compound manifolds are so designed that they may be easily used with a dual pipe or header system. Each of these manifolds exhausts four cylinders. Large diameter exhaust pipe should be used to ensure that no restriction will be encountered by the exhaust gases to rob your engine of precious horsepower.

FLYWHEEL AND CLUTCH

The flywheel can be chopped if you are careful to remove the weight so that the strength of the flywheel is not weakened. The stock diaphragm-type clutch is not satisfactory when used with a hopped-up Buick engine. A ROCKFORD type clutch with heavy springs should be used to hold the additional torque produced by your reworked Buick engine. Be sure to use a new clutch disk. The BORG AND BECK replacement type clutch No. 4050 can be used satisfactorily.

ROCKER ARMS AND PUSH RODS

When you rebuild your engine, give special attention to the rocker arm shafts and bushings. If they are worn they should be replaced.

The oil passages and fittings in both the head and rocker arm shaft should be carefully cleaned to ensure adequate lubrication to the rocker arms. The stock push rods are tubular and have adequate strength for even the most radical racing engines.

Tuning Your Engine

Many an engine is running today that is not even coming close to developing the horsepower of which it is capable. Some of these fall short of maximum horsepower by as much as ten to fifty horses. Anyone will admit that is a tremendous loss! A great number of these engines could be tuned to be winners, if their owners would but learn to carefully complete the job of tuning after the engine has been installed in the chassis.

Merely bolting an engine into place in a chassis is but the very first step in building a real "go-fast" special! If you expect the engine to run perfectly and develop its maximum horsepower the first time that you fire it up, then you are (1) an expert that has built a dozen identical engines and tuned all of them to perfection, or (2) laboring under the false impression that an engine can be expected to tune itself.

Fuel, carburetion, clearances, compression, ignition timing and many other factors will vary from engine to engine, and figures are always approximate as regards jet size, ignition timing and gear ratio, unless the engine is being run with identical equipment and under similar conditions as engines built previously.

When we set up an engine in the California Bill Shop, we know from previous experience the factors and details necessary to make the engine run as best it can. Still, we like to check out an engine on the road after it has been installed in the chassis in which it is to be used.

Better yet, a dynamometer test on a chassis or engine type dynamometer (with the proper test equipment) will indicate the exact jet sizes and spark settings to be used as well as indicating the best fuel for the engine if several are being compared. Either type dynamometer can be used with equal success.

Fuel selection will not play any part in the tuning of your engine if you always use gasoline, and by that we mean filling station "pump" gas.

In these days it is necessary to go all-out to win at drag races, speed trials, or on the track. There is no alternative but to use the finest fuels available if they are allowed, as everyone else against whom you are competing is going to be doing the same thing if they are smart. The use of alcohol plus oxygen bearing fuel additives or high octane aviation gasoline is essential if you wish to be a winner.

In the chapter concerning fuel, the author has presented formulae by which you can mix your own fuel with the assurance that it will give you exciting, winning performance. Once you use alcohol for a race, and get the combination right, you'll not be tempted to race with gasoline again, even though the cost is twice as much with alcohol as with gasoline.

The dynamometer test reports show adequate proof of the superiority of other fuels over gasoline. Many readers will be well acquainted with this fact, but it is obvious from the letters that we receive that Mr. Average Car-Owner knows absolutely nothing about other fuels and the manner in which they must be compounded and used in an engine.

To get the fuel/air ratio correct, we consider it advisable to employ a fuel/air ratio gauge, commonly known as an exhaust gas analyzer. For easiest tuning, the ZENITH carburetors have proven themselves superior to all others for use on the G.M.C., Chevrolet and other motors. These carburetors are rather expensive for the Buicks, and we usually use large Strombergs for Buick installations, as the Zeniths for the Buicks will run you $60 each. For the Chevy and G.M.C.'s, these fine carburetors are only slightly higher priced than Rochesters or Carters.

The main reason that we recommend the ZENITHS is that you can engineer the air flow into the engine properly, as there are many venturi sizes, in steps of 1 millimeter size. Dynamometer tests on our own engines proved that too large or too small a venturi by even as small an amount as one or two millimeters will definitely rob horsepower from the engine.

While the STROMBERG carburetors have many sizes of main jets, you have a choice of only four venturi sizes for the Chevrolet or G.M.C. engine (1-1/32"; 1-3/32"; 1-5/32"; 1-7/32"). If you "happen" to choose the proper engine size and venturi size, two or three Strombergs may work to perfection, but why take that chance? Zeniths are available with a complete range of sizes for power jets, main jets, compensating jets, needles and seats, air bleed jets and idle jets, as well as venturi!

To try to save a little money on carburetion by installing cheap carburetors is to defeat the purpose of tuning your engine for maximum output.

Not only that, but it is quite discouraging to try to tune a dual or triple manifold which does not have a rigid linkage system that is absolutely positive and effective, that gives throttle control on each carburetor equally, opens all in unison, and returns to idle without use of several separate springs or one large strong one to get a proper idle. A good manifold should idle properly each time your foot is removed from the accelerator pedal while using but one stock carburetor returning spring (For boats, track, or dry lakes operation we usually increase the throttle return tension as a matter of safety first, as excessive spring pressure on the throttle for short competition use is not objectional and regarded by some as a safety measure, especially for boat racing in choppy water or dirt track racing in roadsters or modified stocks.

Chevrolet hi-torque engine installed in Crackerbox Class boat owned by Bob Palmini. Boat is quite fast and accelerates like a bomb! Used for towing skiers as well as racing.

WHEN INSTALLING A DUAL MANIFOLD OF ANY TYPE:

1. Remove old manifold, scraping all head flanges free of old gaskets;
2. Remove all fittings from old manifold, to be used on new installation;
3. Install carburetors on manifold with linkage;
4. Synchronize the throttles to closed position while the manifold is still on the bench;
5. Set throttles so engine will idle after starting.
6. If using a used manifold, check to see that all flanges are flat and in the same plane. This check can be made by laying a straight edge or long square across the flanges.
7. Use a complete set of new gaskets for the installation. This is very important.
8. Bolt dual manifold loosely to intake ports of head;
9. After making sure that all bosses and flanges are properly lined up, tighten the manifold securely. A misaligned manifold may cause broken flanges when it is tightened down;
10. If using a heated manifold, tighten bolts to exhaust manifold last. Use a new gasket on the heat riser;
11. Shorten or lengthen throttle rod to hook into manifold linkage so that full throttle is obtainable and carburetors return to idle position easily and fully when pressure is released from the throttle.
12. A misalignment of carburetors or linkage will cause binding of throttle shafts and this will keep carburetors from returning properly to idle position. Sticking accelerator pumps will also cause the throttle to fail to return to its idle position properly.
13. Oil linkage joints thoroughly;
14. Shorten or lengthen fuel lines and vacuum lines as required.
15. Make sure inside of fuel line is de-burred so that metal fragments will not get into carburetor and clog needles and seats;
16. Turn the engine over with the starter before connecting the fuel line to carburetors. This will flush out any dirt or foreign particles from the end of the fuel line;
17. Attach fuel lines to the carburetors;
18. Connect vacuum advance line to front carburetors;
19. Start engine.
20. Warm up to normal operating temperature;
21. Synchronize carburetors by using a NICSON Synchro-Gage. It is possible to synchronize the throttles by placing a piece of windshield wiper hose, first on the throttle inside carburetor, and then on the other one, while listening at the opposite end of the hose. The carburetors should be adjusted until each makes the same sucking noise through the hose;
22. Attach a vacuum gauge to the manifold. Make certain that this is hooked directly into the manifold and not to a vacuum booster pump;
23. Adjust the air bleed idle screws until maximum vacuum shows on the gauge;
24. Adjust the idle speed screws to obtain 450-500 RPM, or approximately 10 MPH in high gear. If a racing cam is being used in the engine, then the idle must usually be set at 600-650 RPM;
25. Check vacuum spark advance to see if it is operating properly. The distributor should move when the engine speed is raised slightly;
26. Re-time engine from six to eight degrees more advance, or so that engine gives a slight ping when accelerating from thirty miles per hour in high gear with full throttle.

IMPROPER FUNCTIONING

1. A flat spot off the idle is caused by main jets which are too lean or by using carburetors which have venturi area too large for the engine. A retarded spark (late timing) will also cause a flat spot off of idle.
2. A miss at high speed under full throttle, such as when winding the engine tight in second gear, may be due to worn out spark plugs or spark plugs which have a heat range too hot for the engine;
3. Check the spark plugs. If they are blistering, install a set of colder plugs;
4. If this does not correct a high speed miss, the distributor shaft should be checked for excessive play, as this will cause the points to float at high engine RPM;
5. The wrong setting of the breaker points within the distributor, or spark plugs with too wide a gap, will also cause a high speed miss.

IMPROPER IDLING
Improper idling may be caused by valves set with not enough clearance, by a manifold leak, or carburetors out of synchronization. Read TROUBLE SHOOTING section.

CARBURETORS
If using ROCHESTER carburetors, install #52 main jets. When CARTERS are used, make, sure that the metering rods are set at exactly the same position. This can be checked with a metering rod gauge. BXOV-2 STROMBERGS can be used as they are; stock.

For best results, ZENITH 10157 carburetors should be used! Settings for these carburetors are as follows for a Chevrolet engine, stock or racing, using gasoline.

DUAL MANIFOLD		TRIPLE MANIFOLD	
Venturi	26		26
Idle	13		13
Power	13 (see note)		13 (see note)
Main	24 (in front carb.)		23 (in each carb.)
	23 (in rear carb.)		

NOTE: Use No. 12 power jets for economy. A triple manifold with ZENITHS should be equipped with No. 14 power jets for flat out, full throttle racing operation.

It is very unfortunate that most manufacturers of dual manifolds do not feel the necessity for giving accurate and adequate information for tuning their manifolds after they have been installed. Not only that, but it has only been quite recently that decent linkages have become available. There are still several manifolds today which have linkages which are so full of sloppy motion and free play that to get the set up running right is almost an impossibility.

If you have a manifold which is equipped with a sloppy, stretchy linkage, then fix the linkage before trying to do any further tuning on your carburetion!

If you are planning to buy a manifold, the first point to check is the linkage. If it is readily adaptable to carburetors designed for the Chevrolet engine, then investigate further. Are there enough holes of the proper size machined into the manifold so that the vacuum-operated accessories can be easily hooked up? Is the manifold machined so that you install and remove it without resort to special tools? (The author has seen some manifolds that were machined so that a socket or box wrench could not be used to tighten the manifold to the heat riser' section of the exhaust manifold!)

One or two manifolds are available which use V-8 carburetors. Before purchasing one of these, figure out a manual spark control, because the vacuum advance mechanism of your Chevrolet distributor can't be used with a V-8 carburetor, and it will not work from a fitting placed directly into the manifold.

ROAD TUNING

When your engine is running pretty close to what you feel is right, then it is time to get out to a long smooth stretch of road and do some really accurate and careful speed tuning. It is a good idea to take along a couple of five gallon fuel cans to place by the side of the road ¼ mile apart to serve as markers. These will serve as reference points for checking the performance of your car. Naturally, your steering, brakes, shock absorbers, and tires should be in A-1 condition before attempting to check your car for maximum speed and acceleration.

A notebook and pencil should be taken along to record changes and results as you tune. Otherwise, you will forget them! Borrow a stop watch if you don't already have one.

If you are tuning for the first time, it is advisable to tune for maximum acceleration through the gears as this will bring you very close to the proper set up for maximum top speed. Your car should be equipped with a tachometer if you are seriously interested in getting the engine to run at its best. A rolling start from a predetermined speed in low gear should be used. As you come abreast of your first marker, start accelerating as fast as possible. You will be at the top of second gear or already in high when you pass the finish marker. If you are carrying a helper to man the stop watch, remember that he will slow down your final times by a second or more due to his additional weight. The weight of one man can be a deciding factor in drag racing —remember that!

As you pass your second marker, declutch and turn off the ignition. Brake to a stop. Check the spark plugs immediately! Refer to the accompanying chart to see how close your mixture is to being correct! The color and condition of the porcelain and base of a spark plug will be your quickest guide to getting your mixture right. Make jet changes and continue tuning until the proper color and condition appears.

If your engine will slightly "rattle" or ping on violent acceleration from 20 mph in high gear, it is probably within a few degrees of being right. Change the spark setting in steps of three degrees, using the octane selector (or calibrated clamp) which is stock Chevrolet equipment. If performance improves, make further changes in the same direction. If it falls off, then go back the other way. Final changes should be made one degree at a time until you get the best performance.

Granted that this procedure takes time, it is nevertheless really important if you wish to get your engine running right. At the same time you can make a few runs and change your shift points. That is, shift from low-to-second at a predetermined speed one time and note the time obtained, and do the same the next time. When the best low-to-second shift point is obtained, then check for the best second-to-high shift point. These are important, and many mark their speedos with arrows to indicate the proper shift points. The Chevrolet gear box is exceptionally touchy on shift points since the ratios are so far apart. Our own '41 coupe falls off almost 2000 RPM when shifting from low-to-second or second-to-high. This causes a noticeable loss in horsepower when the engine RPM's fall off by this amount. If you are really going all out to get the best performance, then refer to the Gear Chapter and see how easy it is to install Lincoln Zephyr gears.

Top speed tuning is undertaken in the same way. However, top speed may require slightly more or less spark advance than that which gives best acceleration. The mixture will remain almost the same.

MIXTURE	CHARACTERISTICS
RICH	SOOTY OR WET PLUG BASES DARK EXHAUST VALVES
CORRECT	LIGHT BROWN COLOR ON PORCELAINS, EXHAUST VALVES RED-BROWN CLAY COLOR. BASE OF PLUG SLIGHTLY SOOTY, NOT TOO DRY. (SHOULD LEAVE SLIGHT SOOT MARK ON HAND WHEN PLUG BASE IS TURNED ON PALM.)
LEAN	PLUG BASE ASH GREY. GLAZED BROWN APPEARANCE OF PORCELAIN (MAY ALSO INDICATE THAT PLUGS ARE TOO HOT).

Exhaust Systems

A chain is no stronger than its weakest link! This old adage certainly holds true with the exhaust system of your Chevrolet, GMC, or Buick. A choked-off, stock exhaust system will rob your engine of as much as twenty five horsepower at top speeds. You realize that twenty five horses cost much money to obtain, so don't keep them bottled up where they can't do you any good!

In order to effectively reduce the back pressure on your engine, the exhaust should be allowed to escape directly into the air from large diameter exhaust pipes. Unfortunately, this is illegal! Except for racing, we can't use such a set up on our cars. Most owners of hot engines equip them with dual mufflers and headers. This improvement on a stock engine will add over ten horsepower at the rear wheels. Think of what it does for a big racing engine!

Even a single steel-packed muffler, if properly constructed, will add four or more horses at the rear wheels. We recommend DOUGLASS dual sets and steel-packed mufflers.

You can make a set of headers for your engine and include header plugs whereby the exhaust can be channeled through the mufflers for normal use, or "un-corked" for racing purposes. The easiest and neatest way to install headers on a Chevrolet is to purchase and install a set of CLARK headers. A set of header plugs can be grafted onto your exhaust pipes.

If installing headers on any of these three engines by using flanges cut out to fit the ports, then it is absolutely necessary to keep the flanges bolted tightly to the head while constructing the headers. After the job is finished, it should be normalized by heating the entire assembly so that the headers will fit on and off of the engine easily. Care must be taken to insure that the intake manifold will clear your headers. Careful attention will be needed to make sure that the headers clear all working parts and pieces in the vicinity of the header.

No part of the exhaust pipes or headers should come into direct contact with the body or frame of the car, or a drumming noise will be the result.

A header plug can be welded into your stock exhaust pipe ahead of the muffler. This can be closed with a standard pipe cap. We usually make these pipes with at least 2" inside diameter. If this pipe is pointed downward, it will collect some water from the exhaust gases, so be careful when removing the header plug, or you'll get a faceful.

Remember to keep the exhaust away from your fuel tank. If running with alcohol or nitro-methane fuels, it is imperative that the header be led completely past the driver's seat and out to one side or the rear of the car. These fumes may not kill you but you'll wish you were dead!

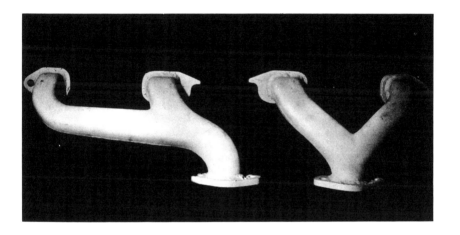

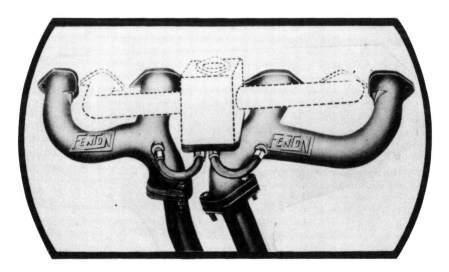

Upper photo shows CLARK fabricated metal headers. Lower photo shows FENTON cast-iron headers. Models for both GMC and Chevrolet are available in each make.

Fuel

Naturally, most of the readers of this manual will never be interested in using any fuel other than gasoline in their Chevrolet or G. M. C. engines.

A few will be eager to learn what can be done in the way of fuel to increase the horsepower available from their engines.

Recent developments in fuel have been largely due to the use of oxygen-bearing fuels, known as mono-propellants; that is, they need no oxygen to complete their cycle of combustion. When these are added to alcohol, (Methanol is the most commonly used racing fuel) or to gasoline, great things occur. Horsepower increases of 10 to 15% are quite common, with more having been obtained in some cases.

The best known and most widely used of these expensive fuel additions is Nitro-Methane. It is not available everywhere in the United States, but we believe that it will not be long until it is. In the course of six months here in Southern California, the data on the use of Nitro-Methane has changed from a whispered rumor to a fact known and freely talked of by everyone involved in racing.

In any Chevrolet engine using alcohol fuel and compression of 10:1 or less, the use of a 20% nitrate mixture is quite common for use in all types of racing. This is obtained by adding a quart of Nitro-Methane to a gallon of Methanol.

Since these "super" fuels produce a far greater amount of B.T.U.'s (heat energy), it is necessary to use larger jets. Usually a 20% increase in jet size is sufficient, but this will vary with the amount of nitrate used, atmospheric conditions, and other factors. Jet changes for nitrates must always be worked out on the individual engine.

Another point to remember is that colder spark plugs must be used to keep the plugs from being burned up. If you use a nitrated fuel with larger jets, but retain your spark plugs which are just right for Methanol, then your plugs will turn a blue-white color, or perhaps even fry the electrode and produce "cackle-berries" (blisters) on the porcelain of the center electrode of each plug.

After fuel has been nitrated, it should be used up as soon as possible. Do not attempt to store it in ordinary cans; a 5% Nitrate Alcohol mixture will eat through the solder in the average 5-gallon can in about one week. Most fellows remove the fuel from their machines immediately after racing is over for the day. Then, straight, un-nitrated fuel is run through the engine for several minutes in order to clean out the carburetor. Always drain your tank, fuel lines and carburetors after a meet.

When using nitrated fuel for the first time, it is usually wise not to exceed a 10% mixture, even though mixtures of 25% have been used successfully.

It is imperative that the jet sizes be increased greatly, otherwise the nitrate will eat up the pistons in a hurry. Plugs should be two to three steps colder than those used for gasoline on the street. (We usually use J-6's on the street, and use J-2 Champions for competition).

Don't just put this fuel in and run it without being willing to experiment to get the correct jet size, or you'll find yourself with a lot of holes where the crowns of your pistons used to be, and that is not idle chatter! Put some richer jets in to start with!

Just get on the throttle about half way in low gear, don't take the engine to peak revs! Then shut off the ignition and clutch out the engine so that you coast to a stop. Check two or three plugs and compare with the plug chart to see how your mixture is burning. If it is just a bit rich that is fine! If too lean, then don't run any more until you drill the jets out larger or install larger size jets. When the mixture appears to be burning right, then get a full load, full throttle reading by taking the engine up to peak revs. Notice the extra acceleration this stuff gives you. Terrific, isn't it? Jerk another plug and check the mixture again to make sure that it is burning all right. If it is, you are ready to go out and "shut-off" your competitors.

Alcohol is sometimes balky to start with a cold engine. In this case it is wise to prime the carburetors with gasoline to get the engine started.

Flooding your engine with alcohol fuel may wash off all of the oil from the cylinder walls and cause a failure to start by losing all of the compression. In this case, the spark plugs should be removed and a small quantity of motor oil squirted into each cylinder. The engine should then be turned over several times before attempting a new start. If the spark plugs that you use for racing with alcohol fuel keep getting fouled up when starting from cold, switch to plugs which are one or two steps hotter in order to start. Switch back to running plugs after engine has reached operating temperature.

Check your oil after running with alcohol fuel. It is usually a good idea to change it fairly often when running alky, as this fuel tends to dilute the oil.

When using alcohol fuels, it is absolutely necessary to modify your carburetors so that they will pass the extra volume of fuel consumed in burning this type of fuel.

The primary changes necessary in converting your carburetors are increasing the size of the main jet, dump tube, and needle valves. Many builders use STROMBERG carburetors when running alcohol due to the simplicity of converting these carburetors to utilize this type of fuel.

When racing enthusiasts first began to use alcohol here in Southern California (before the World War), many felt that the alcohol fuel would not be good for the engines and that it would cause rapid deterioration of the engine parts. Experimentation with many engines has proven to us that alcohol fuel is beyond a doubt easier on an engine as the carbon formed by gasoline fuel is almost non-existant with alky, and the alky burns nice and cool, keeping internal engine temperatures down to a minimum. Not only that, but alcohol can be used with higher compression without danger of detonation.

Apparently the highest compression ratio practical when using gasoline is approximately 10:1, whereas alky can be used successfully with compression ratios as high as 16:1, although we do not recommend that you use compression ratios over 10:1 when using the stock reworked Chevrolet Head. However, you can use up to 12:1 with the G.M.C. head without detonation trouble, due to the advanced design of the combustion chamber.

HORSEPOWER vs. COST

Fuel comparison or price with power produced. Prices are for one gallon of fuel.

Fuel	H.P.	Price
Ethyl Pump Gasoline	140 H.P.	27¢
140 Octane Aviation gas	150 H.P.	$1.42
10% Nitromethane and 10% Amyl-Acetate		
Methanol	175 H.P.	75¢
Methanol with 20% Nitro	195 H.P.	$1.75

Water Injection

Any internal combustion engine runs more smoothly on a foggy or rainy day. You have noticed this yourself. This is because water-vapor mixes with the incoming fuel and slows its burning to give more complete combustion throughout the combustion chamber. This reduces spark-knock or detonation to a marked degree.

Water-vapor with your fuel will permit you to utilize more spark advance with any compression ratio. It will also permit the use of regular type fuels with high compression engines. Some cars are designed so that they will receive an over-rich fuel mixture at top speeds, or under full throttle loads. Water-vapor can be used instead to give a cooling effect for full throttle operation.

Perhaps the most fabulous feature of water injection is that it will completely eliminate carbon from the inside of an engine. It will seldom produce any more horsepower or fuel mileage, but the fact that cheaper fuels can be used to get the same performance, that the engine is kept clean, and that full use can be made of your high compression performance, will make one of these units well worth your money.

The latest Water-Injection unit on the market is the OCTA-GANE. It is designed so that vacuum from the intake manifold, and pressure from the exhaust manifold, are used to control the injection of water-vapor into the engine. These units are available to fit single or multiple installations. Photos and drawings of this unit's operation are presented here. These are superior to the older water carburetors formerly manufactured by the same firm. The author recommends them after checking with the rules presented in his last manual: (1) Water is injected when it is needed, at full throttle and under high load conditions; (2) No water is bled into the engine at idle or cruising speed; (3) There is no possibility of water being siphoned into the engine after the engine has been shut off.

Spalding Brothers' (Tom and Bill) Chevrolet powered track and dry-lakes roadster. Torsion bar equipped. Using Torsion bars by John Hartman in conjunction with special English-made ROTO-FLO shock absorbers, this car proved to be a masterpiece in handling. Engine was of course equipped with Spalding cam and ignition. Fastest dry lakes time was 149 mph. Engine produced 248 horsepower from 248 cubic inches, using the famed Hilborn injection system.

John Hartman's lakes roadster has been clocked at speeds in excess of 150 mph. This car was winner of the second "Little 500" Memorial Day race event held at Carrell Speedway

Driving Your Hopped Up Car

It would be impossible to give you, our reader, a completely detailed course in driving a racing car. Even so, a few brief pointers may not be out of line at this time.

While the author is no preacher, he would still like to write down a few ideas which may be helpful. Well meaning citizens who see a high performance car laying off rubber through the gears may not recognize the beauty of your carefully assembled machinery working in perfect coordination with your flawless driving skill. Instead, they may arouse an entire area against you and your "hot-rod," as well as any other fast cars in that area.

After you have built your fast Chevrolet, Buick or G.M.C., take it easy! This especially applies around town and built up areas. Don't show off the capabilities of your machine where it could cause unkind and undue comments about you and your machine.

Resist the temptation to drag race from stop lights or stop signs. Drag racing is lots of fun! We certainly don't deny that! We couldn't, because the author is a dyed-in-the-wool drag racer, if there ever was one! Restrict your dragging activities to legalized drag strips or to long straight roads with perfect visibility and little traffic. There are usually a few like this in every locality where an occasional race will go unnoticed by the local gendarmes.

Your hot engine will make your car a bit more tricky to handle, but it is a pleasant feeling to have to "get off" the throttle a little to help straighten out on a city corner. You will have a lot more power at the rear wheels which will tend to lessen traction when full power is applied. Thus, it will be necessary to take it easy where the streets are messy or the surface is loose.

The sport of owning a high performance car is comparable to none other. It is great! Don't spoil it for others by being a show-off or a hooligan. One or two unthinking fellows can spoil things for miles around. For details on clubs and how to form them, we suggest that you read all of the leading speed magazines regularly.

To protect yourself and others, and have a lot of fun, it is a good idea to get the full cooperation of your local police department. Authorized drags and speed trials can be set up easily and have been legalized already in many areas of the U. S. The spectator interest is terrific

Trouble

ENGINE TROUBLE

No matter how constantly you attend to the servicing of your engine, or adjust it, trouble will sometimes occur in the way of a breakdown on the road or failure to start.

The author makes no claim to having formulated this listing and frankly admits that it has been compiled from several automotive books contained within his library. Anyone is liable to have trouble with his engine, be it stock or hopped up! Having had his share, with no check list to go by in many cases, the writer trusts that these listings and explanations may save his readers a large amount of time and frustration, and that they may thereby derive more pleasure from the use of their engines.

DETERMINING THE TROUBLE

The difficulty of stoppage can be usually laid to one of three types of trouble:

1. **ELECTRICAL**
   a. Usually sudden stopping of the engine
   b. Seldom gives any warning
2. **CARBURETION**
   a. Engine may spit back or misfire
   b. Engine may start, run erratically, and stop again
   c. Engine usually runs a few seconds or minutes after trouble develops
3. **MECHANICAL**
   a. Usually accompanied by noise if a part has broken or worn excessively
   b. Seldom causes engine to stop completely, although may cause spitting and show a tendency to miss on one or more cylinders

A vacuum gauge and a compression gauge are very useful items for use in finding the source of trouble. We usually hang a vacuum gauge under the hood of any car on which we are performing a tune-up or trouble-tracing operation. Most engine "tune-up" type vacuum gauges are marked to facilitate the location of trouble.

While most engines using racing camshafts will show less than the vacuum which would be normally obtained with the stock cam, it is safe to assume that any good engine should read 15" to 18" at idle. If the gauge reads less, check for:

   a. incorrect valve clearances
   b. leaky cylinder head gasket

c. leaking intake manifold or intake manifold gaskets will cause engine to "hunt" or vary RPM's up and down while idling
   d. this may be checked by squirting lube oil around gaskets in abundance

To check individual cylinders when using the vacuum gauge, short out one spark plug at a time and read the drop shown on the vacuum gauge. Each cylinder, when shorted out, should cause an equal drop in vacuum registered on the gauge. This drop should be about one to three inches of vacuum for each cylinder. If the drop is not uniform for each cylinder, then check for (a) and (b) just mentioned above.

The difficulty may be caused by a valve or valves not properly seating, broken or worn valve springs, or other defects in the valve mechanism. A cracked piston, or one with poor rings, or a bad cylinder wall can also cause less drop on an individual cylinder in which they occur.

If only a small drop shows on the gauge for each cylinder, and this occurs with all cylinders, then check:
   (1) For secondary ignition trouble
   (2) Leak in intake manifold or gaskets
   (3) Choked up exhaust system
   (4) Worn out carburetor, leaking around butterfly or throttle shaft
   (5) Improper idle-air adjustment

POOR STARTING OR FAILURE TO START

1. Battery low, starter robbing all juice from the coil. Check the battery and cables. Terminals may be corroded, cables worn or shorted. Car will usually start by pushing.

2. If battery O.K., turn engine with starter. Hold one spark plug lead ¼" from engine block. A good fat blue spark should occur when engine is turned over and ignition switch is on. A yellow or whitish spark indicates low voltage. If no spark, then check for:

   (a) cracked or wet distributor cap
   (b) worn points or dirty points
   (c) points set incorrectly
   (d) faulty coil, primary or secondary windings
   (e) ignition wire shorted or grounded
   (f) bad condenser

(3) Check carburetor by looking into it to see if working the throttle

causes fuel to squirt into the manifold. If no fuel shows, then check for:
- (a) fuel in tank
- (b) fuel pump, should be gasoline in bowl
- (c) leak in line between fuel pump and tank

4. The next thing to try if everything else checks O.K., is to push the car. If this starts the car, then:
- (a) starter switch contacts may be worn or corroded or failing to make contact
- (b) battery terminals may be corroded enough to prevent sufficient current from reaching the starter

5. Engine may make a feeble attempt to start and not do so, or may kick back.
- (a) spark plug wires may be crossed, or incorrectly on plugs for proper firing order
- (b) engine may be flooded
- (c) if engine flooded, clear excess fuel from cylinders by turning the engine rapidly and holding the throttle wide open at the same time.
- (d) valve clearances may be set so close to reduce noise that the valves do not seat properly and thus cause a lack of compression.

ENGINE FAILS TO START OR WILL NOT RUN AFTER STARTING

If your engine is balky and won't start at all after making modifications to it, then check the following points:

1. Engine may be flooded, or may need a few ounces of raw gasoline squirted into the intake manifold;

2. There may not be any fuel, or perhaps not enough fuel reaching the carburetor;

3. Idle screws may not be set correctly;

4. Starter may not be cranking fast enough to start engine. With a new or stiff engine, we often use two 6-volt batteries connected in series so that the starter receives 12 volts. Coil should not get more than its rated voltage or it may burn out;

5. Intake manifold may be leaking;

6. Intake manifold may be restricted (rag in port or overhanging gasket, etc.);

7. Defective or fouled plug;

8. Defective ignition;

9. Incorrect ignition timing;
10. Incorrect valve timing — timing gears may not be properly meshed;
11. Defective ignition switch;
12. Valve sticking, leaking, or improper clearance;
13. Loss of compression;
14. Loose or shorted primary ignition wires.

ENGINE STOPS SUDDENLY
1. Ignition circuit grounded or broken;
2. Out of fuel;
3. Vapor lock or air in fuel lines;
4. Carburetor jets clogged with dirt;
5. Fuel filter clogged;
6. Sheared timing gear.

ENGINE DOES NOT DEVELOP FULL POWER
A. Carburetion
   1. Too lean or too rich;
   2. Throttles not fully opened;
   3. Incorrect fuel for the engine;
   4. Manifold leaking air;
   5. Low fuel pressure;
   6. Restricted fuel flow;
   7. Choked or clogged air cleaners.
B. Ignition
   1. Spark plugs defective;
   2. Breaker points dirty, worn or pitted;
   3. Weak battery;
   4. Faulty coil;
   5. Cracked distributor cap;
   6. Automatic advance functioning improperly;
   7. Improper ignition timing;
   8. Wrong heat range spark plugs;
   9. Engine clearance may be set too close throughout.
C. Miscellaneous
   1. Detonation;
   2. Incorrect valve timing;
   3. Incorrect valve clearance;
   4. Engine too cold or too hot;
   5. Choked exhaust system;
   6. Leaking valves;
   7. Blown head gasket;

8. Low compression;
9. Weak valve springs;
10. Engine badly carboned up.

ENGINE DOES NOT IDLE PROPERLY
1. Dirt in carburetor;
2. Low compression;
3. Leak in intake manifold;
4. Incorrect valve clearance;
5. Spark plugs foul when idling, due to rings not sealing, spark plugs too cold, gap too narrow;
6. Carburetor throttle shafts or butterflies worn;
7. Idle mixture rich or lean;
8. Bent distributor shaft;
9. Cracked distributor cap;
10. Incorrect valve timing;
11. Water in gasoline;
12. Cold engine;
13. Leaking needle seat;
14. Check valves in carburetors. May be stuck;
15. Bad valve guides;
16. Bad windshield wiper connections. Check under dash also, leaky windshield wiper hose.

Bill Johnson's Chevrolet Race car. This car qualified second fastest in a field of Offies at Arlington, Texas in 1950. Engine produced over 260 horsepower.

G.M.C.'s, HOW MUCH HORSEPOWER?

ENGINE: 274 INCH PROFESSIONAL RACING
Fisher head and pistons, roller-tappet cam, 2 Vacturi floatless carburetors, 12:1 compression, 270 block with 248 crank, methanol fuel/10% nitro-nethane.

285 H.P. at 5000 RPM

ENGINE: 292 INCH FOR DRAG, TRACK, OR LAKES
Howard cam and five carburetor manifold, 302 GMC head, 3-15/16" bore Fisher pistons, large exhaust valves, 270 block and crank, straight methanol fuel.

265 H.P. from 4000 to 5000 RPM

ENGINE: 292 INCH DRAG
Fisher head and pistons, Howard F-6 cam, Spalding ignition, 6 Stromberg 97 carburetors, 2 Bosch Big-Brute coils, methanol fuel/15% nitro-methane.

310 H.P. at 4800 RPM

ENGINE: 326 INCH DRAG OR LAKES
Howard cam and five carburetor manifold, special Fisher pistons, 270 block with stroked 270 crankshaft, 270 H cylinder head with large exhaust valves, 14:1 compression, Scintilla magneto, straight methanol fuel.

281 H.P. at 4700 RPM

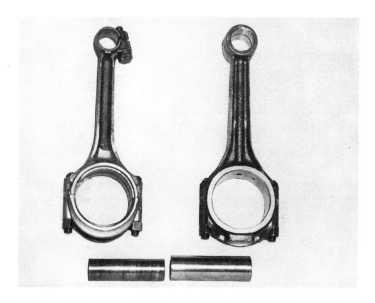

Comparison of Chevrolet and GMC connecting rods and pins. Note ruggedness of the GMC rod and pin. GMC rods can be used in pressure oiled Chevrolets, when the proper pistons are installed to fit the pins.

Don Nicholson's 1950 Chevrolet is equipped with a 270 GMC bored out to 302 cubic inch displacement. Engine features HOWARD M6 cam, McGURK pistons, SPALDING ignition, SUN tachometer, and HOWARD five carb manifold. Turns 125 mph. on pump gas, or 95 mph. in standing quarter-mile drag race. Additional features are 1949 Ford transmission with overdrive, 25 tooth Zephyr gears, and Oldsmobile one-piece wraparound windshield.